PROGRESS IN PLANNING

Editors

D. Diamond & B. H. Massam

PERGAMON

Aims and Scope

To provide a convenient and rapid outlet for extended papers drawn from recent social science research of importance for the practice of urban and regional planning and environmental management. New and innovatory aspects of planning are emphasized: the contributors are international in character and material is drawn from the entire range of social sciences throughout the world.

The range of disciplinary interests covered include: sociology (especially urban sociology), demography, development planning, environmental studies, economic, social, urban, rural and population geography, civic design, transportation planning, political science and public administration, social administration and urban and regional economics. The methodology of planning, the philosophy and purpose of planning and its relationship to long-term forecasting, the management of information systems and social theory are also included.

Progress in Planning covers the following subject categories:

Transport Planning
Planning Theory and Method
Regional and Economic Planning
Third World Planning
Industry and Employment
Physical, Urban and Environmental Planning

The Item-fee Code for this publication is: 0305-9006/98 $19.00+0.00

∞™ The paper used in this publication meets the minimum requirements of American National Standard for Information Sciences—Permanence of Paper for Printed Library Materials, ANSI Z39.48-1984.

Transferred to digital printing 2006
Printed and bound by Antony Rowe Ltd, Eastbourne

Meaningful Collective Choice? Public Planning and Arrow's Theorem

TORE SAGER

Department of Transportation Engineering, Norwegian University of Science and Technology, 7034 Trondheim, Norway

PERGAMON

Biography

Tore Sager is Professor in the Department of Transportation Engineering, Norwegian University of Science and Technology, Trondheim, Norway. He graduated as an economist and received his Dr.techn. degree at the Nordic Institute for Studies in Urban and Regional Planning, Stockholm, in 1990. Major fields of work have been transport economics, urban land use, transportation planning, and techniques and methods in planning. His current research interest is planning theory, particularly communicative planning theory and the study of bounded rationality in transportation planning processes. He published *Communicative Planning Theory* (Avebury) in 1994.

Contents

Abstract 75

1. Introduction 77

2. The Paradox of Social Choice 81

2.1. Arrow's Impossibility Theorem 81

2.2. Exogenous Variables in Planning and Social Choice 86

2.3. The Planning Process as an Arrow Type Aggregation Device 88

3. Synoptic Planning as Relaxation of the Arrow Conditions 93

3.1. Social Choice Implications of the Synoptic Planning Phases 94

3.2. The Impact Assessment Paradox 96

3.3. Synoptic Procedures Weakening Logicality 100

4. Incrementalism as Relaxation of the Arrow Conditions 103

4.1. Disjointed Incrementalism as a Procedure Relaxing the Condition of Unrestricted Scope 103

4.2. Properties of Disjointed Incrementalism Modifying the Independence Condition 107

4.3. Comparing Synoptic and Incremental Planning as Strategies for Alleviating the Effects of the Social Choice Paradox 111

5. Social Choice and Communicative Planning 115

5.1. Introduction on Rationality and Power 115

5.2. Dialogue and Social Choice 117

5.3. The Arrow Conditions and Conflict Management 120

6. How Decision Cycles Arise: Two Cases 123

6.1. The Parliamentary Vote on the Siting of the New Main Airport in Eastern Norway 123

6.2. The Tragic Tale of the Trondheim Tram 127

7. Conclusions 135

Bibliography 139

Pergamon

Progress in Planning Vol. 50, pp. 75–140, 1998

Printed in Great Britain
0305-9006/98 $32.00 + 0.00

PII: S0305-9006(98)00014-2

Abstract

This essay is based on Kenneth Arrow's celebrated impossibility theorem stating that under a few quite reasonable assumptions, collective choice cannot be simultaneously logical and fair. The purpose here is to argue that planning procedures can reduce the likelihood that decision cycles will arise when democracy is pursued.

It is examined whether some of Arrow's assumptions can be relaxed under widespread and participatory planning, since planning and public debate may force a minimum of conformity on the stated individual preferences. However, collective choices have to be made in spite of the impossibility theorem. Well-known theories of planning, especially synoptic planning and disjointed incrementalism, are analyzed to assess if they are acceptable ways of organizing decision-making processes in the face of Arrow's impossibilities. Some aspects of the planning modes accept or imply that logicality is deficient and that cycles may occur, while other facets arrange for less than perfect democracy and fairness.

The social choice implications of evaluation technique and communicative forms of conflict management are analyzed, and two transportation planning cases are studied in order to show how decision cycles arise.

CHAPTER 1

Introduction

This essay is closely bound up with Kenneth Arrow's (1963) celebrated social welfare theorem (first presented in 1950). The theorem demonstrates, under apparently plausible assumptions concerning the aggregation of individual preferences, that no suitable social welfare function can exist. Suitability would require that collective choice be both fair from a democratic point of view and logical. But we are unable to define an acceptable amalgamation device under the nearly omnipresent condition of conflicting interests. Since no decision process observes Arrow's conditions of simultaneous logicality and fairness, the possible processes display various trade-offs between these desired properties.

Even if it is impossible to make collective decisions that simultaneously satisfy the claims to logicality and democracy, actual governments have to make social choices. This implies that they relax one or more of the conditions giving rise to Arrow's paradox of social choice (Douglas, 1984). It is done by political and bureaucratic procedures, sometimes requiring specific forms of communication and sometimes assisted by analytic techniques—and planning is politics, communication, and technique.

The impossibility theorems of social choice are of interest to planners for several reasons. Firstly, there is the question of rationality. The theme of this exposition seems to be especially salient to a decision-centred view of planning, which is shown by Faludi (1987) to be primarily concerned with the rationality of ongoing action. Is public planning condemned to prepare for collective choice that is either indeterminate or unfair? If so, why should planners make such strong efforts to be rational? And what should be meant by rational planning in these circumstances? Faludi (1986, p. 116) specifies his critical–rationalist planning methodology as a theory of how we can justify our decisions. I do not find such a theory sufficiently complete unless it explains how planning can meet the challenge posed by social choice theory. Given Arrow's social choice paradox, one important sense in which planning can be rational is by specifying procedures implying the relaxation of the Arrow conditions in a manner causing the least possible unfairness or the least possible politically harmful decision cycles.

Secondly, it may be possible to show that the set of alternatives and the individual preference rankings are often changed throughout the planning process in such a way that the final political decision can be fair and meaningful. If so, this is an argument in favour of public planning that is usually overlooked.

Thirdly, a number of planning techniques are closely related to decision-making. They can therefore properly be assessed by the criteria applied by Arrow when establishing his impossibility theorem. I refer primarily to formal evaluation techniques and conflict management techniques.

Fourthly, the planning process can be similar to a decision process. The final decision on which planning alternative is the 'best' one is often made by an assembly of elected politicians. However, some contend that this is merely rubber-stamping, or that more interest should be placed on the preceding policy-making processes where the politicians sometimes do and sometimes do not participate and exert control. In both cases it would be appropriate to conceive of the entire planning process as a mechanism for making collective choice. When this is approved it becomes interesting to compare the synoptic and incremental planning modes with the qualities required by Arrow. The article explores which relaxations of the assumptions leading to Arrow's impossibility theorem are implied by synoptic and incremental procedures. I intend to show that the two planning modes constitute alternative strategies, not to dissolve the paradox, but to proceed towards collective decisions in the face of paradox.

The reasonableness of the Arrow conditions to be spelled out in the next chapter is discussed in, for example, Arrow (1963), Plott (1976), MacKay (1980), and Riker (1982). To judge from everyday practice in politics and planning—where interpersonal comparisons of welfare are made all the time—it would seem judicious to allow such comparisons in a collective decision device. However, all the authors just mentioned, and most other social choice theorists, find that interpersonal comparison of utility has no clear meaning, and so they rely on information about individual preference rankings only. What if you—for this or other reasons—conclude that the assumptions leading to the paradoxes of social choice are unrealistically restrictive? The three points below might make you consider to go on reading anyway.

(1) It is well demonstrated that attractive features of social decision procedures must be renounced when violating the Arrow conditions in various ways. For instance, insisting on interpersonal comparisons of utility introduces arbitrariness in the collective ranking and weakens the fence between the social choice and the process leading up to it, thus encouraging manipulation (MacKay, 1980, Ch. 4).
(2) It might be of some interest to examine how public planning can help to prevent social choice from being either unfair or illogical, even if you believe that this does not *have* to be the characteristics of the collective choice mechanism without planning.
(3) You might find it interesting to explore the features of public planning regarded as a decision-making process and see what values are strengthened when the procedural planning model is brought more in line with the Arrow conditions.

Central results in social choice theory suggest frequent and extensive decision cycles (McKelvey, 1976; Riker, 1982). This is not what is observed in practice under democratic regimes, however. The striking contrast between theory and practice is given much attention in the social choice literature (Green and Shapiro, 1994, Ch. 6). A number of explanations have been proposed, and the notion of structurally-induced equilibrium takes a prominent position (Shepsle, 1979; Shepsle and Weingast, 1981). Public planning is an institution that can help to provide the structure needed.

The essay is arranged as follows. The second chapter presents Arrow's impossibility theorem and explicates the underlying assumptions. It is argued that at least part of the planning process (the evaluation phase) forms a close analogy to Arrow's problem of aggregating individual preferences. Chapters 3 and 4 discuss well known planning theories in a social choice perspective. It is explained how properties of synoptic planning violate the Arrow conditions and modify logicality by accepting information on preference intensities. Disjointed incrementalism relaxes two of Arrow's conditions, and the most peculiar consequence is that fairness is modified. The fifth chapter links Arrow's problem to communicative rationality. It is explained how dialogical planning and conflict management can make it reasonable to relax some of the Arrow conditions and facilitate transitive and fair collective decisions. Part of this chapter shifts the focus from theories of planning to particular planning techniques. The sixth chapter present is two case studies of indeterminate policy outcome in the transport sector. It is shown how new issues, new alternatives, and shifting political tactics can change coalitions and lead to intransitivity in the policy-making process. The last chapter sums up and offers some conclusions.

CHAPTER 2

The paradox of social choice

The chapter presents Kenneth Arrow's social choice problem. The explicitly stated conditions for his theorem are first delineated, then I discuss some implicit presuppositions concerning what can be considered as exogenously given. For instance, the individual preference orderings are accepted as given facts when Arrow formulates the problem of social choice. I explain how decision-making in the planning process differs from Arrow's model. It is suggested that public planning has a bearing on social choice in two ways. Firstly, because of deliberation throughout the planning process can affect the individual preferences before they are aggregated. Secondly, because a number of small decisions are made in the various planning phases, which affect the agenda and in turn have an influence on the final political decisions. The planning process can therefore be studied as a social choice process.

As long as unanimity reigns, collective choice is easy and straightforward. When individuals disagree, some kind of aggregation of their preferences has to take place before a collective choice is made. Majority voting is one such amalgamation procedure, and it has long been known that it may result in intransitivity instead of a Condorcet winner.[1] Under certain conditions there will be a majority behind each of the several alternatives in pairwise comparisons, and this is referred to as the paradox of voting. An example is found in Section 6.1.

Kenneth Arrow set out to investigate if there could be an aggregation technique evading intransitivity, yet fair and compatible with democracy. His negative result is called the general impossibility theorem of social choice (Arrow, 1963). It is affiliated with the voting paradox but has consequences far beyond that well known problem (Sen, 1985). Both paradoxes point at important difficulties of democratic government.

2.1. ARROW'S IMPOSSIBILITY THEOREM

Drawing on MacKay's (1980) exposition, I will now indicate what is implied by the conditions yielding the paradox of social choice. Arrow proved that no aggregation device can jointly satisfy the following four conditions and produce transitive collective preference orderings:[2]

(1) (U) Unrestricted scope: An acceptable device should be able to process any (logically) coherent set of individual preference rankings of any number of choice alternatives.

(2) (P) The Pareto principle: When every individual without exception prefers X to Y, the device must rank X above Y in its social ordering. That is, an acceptable device must honour unanimity.
(3) (D) Nondictatorship: An acceptable device is not to be dictatorial. The aggregation device is said to be dictatorial if there exists a person so that the social preference ordering is always the same as the individual preference ordering of this person.
(4) (I) Independence of irrelevant alternatives: The social ordering of a given set of alternatives should depend only on each individual's ranking of those alternatives.

William Riker (1982, p. 132) explains why transitivity—or some less restrictive condition ordering the collective preferences—is attractive.

> (T)hey often forestall manipulation by some participants either of agenda or of sets of alternatives to obtain outcomes advantageous to the manipulator. As Arrow remarked at the conclusion of the revised edition of *Social Choice and Individual Values*, 'the importance of the transitivity condition' involves 'the independence of the final choice from the path to it'. 'Transitivity', he said, "will ensure this independence thereby ensuring also that the preferences of the participants (rather than the form of or manipulation of the social choice mechanism) determine the outcome.'[3]

Turning now to the conditions (1)–(4), they are not all equally attractive candidates for relaxation in a society aspiring to democracy. I will assume that unanimity is honoured, leaving the Pareto principle (P) untouched. One could even claim that planning is often justified by its ability to realize Pareto optimality by correcting other allocation mechanisms. For example, a perfectly competitive market will produce an outcome that is not Pareto optimal when public goods exist. There is often a higher level of production of the public good and another distribution of the corresponding costs, that would be preferred by everybody. Uncoordinated decisions made by different agencies sometimes give results that everyone would want to alter (Johansen, 1979)—compare public agencies digging up the same street by turns. Similarly, dictatorship is so contrary to the democratic ethos that condition (D) is considered valid. Some remarks on the concentration of power are offered later in the chapter. In a democracy it seems more appropriate to strengthen than relax the condition of nondictatorship, e.g., by demanding that oligarchies not be allowed.[4]

So, to support the contention that actual governments relax some of the Arrow conditions in order to make social welfare judgements which depend on the preferences of all individuals, we are left with unrestricted scope (U) and independence of irrelevant alternatives (I). Before I discuss how they are both modified by well-known modes of planning, I should spell out what is meant by (U) and (I), and suggest general ways to relax them.

Unrestricted scope means that any finite number of choice alternatives is permitted. A reduction to two alternatives evades the paradoxes of both majority voting and social choice. Pairwise comparisons may therefore be part of the strategy for making collective decisions. Riker (1982, pp. 121–22) shows that Arrow's theorem gains in practical relevance with increasing number of alternatives and decision-makers. Restrictions on these variables reduce the likelihood of preference cycles. Thus, agenda setting is an important task in the decision process. A recommendation often met in theoretical work on citizen participation is that each significant interest group should find its interests taken care of in at least one of the planning alternatives being evaluated. This proposition can increase the risk of cycles in a complicated planning case affecting many interests.

Unrestricted scope also means that any transitive and connected preference ordering of the choice alternatives should be accepted, and any combination of such individual orderings should be allowed. Each individual is free to rank the alternatives in any logical way permitted by the preference and indifference relations. One extreme situation occurs when everybody has the same top alternative. The Pareto principle then assumes that this alternative is the social choice. However, agreement in this sense is not necessary to prevent preference cycles. A much weaker kind of agreement is required by single-peakedness, perhaps the best known of the value restrictions ensuring transitivity of the collective preferences (Black, 1958). A single-peaked curve may be (1) always downward sloping, (2) always upward sloping, (3) sloping upward to a particular point and then sloping downward, (4) sloping upward to a plateau and then sloping downward, (5) horizontal and then downward sloping, (6) upward sloping and then horizontal. A set of preference orders, one for each member of society, is single-peaked if some ordering of the alternatives on the horizontal axis allows every individual preference ordering to be drawn as a single-peaked curve (Riker, 1982, p. 124). This more modest agreement, on an ordering criterion rather than on ranking, has as a consequence the barring of cyclical majority patterns.[5]

Single-peakedness implies that the alternatives can be lined up along a common scale or dimension. Each voter would then rank the alternatives lower with increasing distance from his favourite (or favourites, when indifference is permitted at the top, as in the above definition). Single-peakedness means that the voters have a common view of 'the situation' in which the plan at hand is embedded, although they may differ widely in their judgement of the quality of the plan. Imagine, for example, that a by-pass is being planned. The alignment may penetrate the outskirts of the town, or it may run further from the centre and traverse mainly farmland but require an expensive bridge. There may be diverging views of what is the main trade-off. Some will regard it as an option between construction costs and traffic safety; others may see traffic noise and preservation of cultivated areas as forming the end points of the scale. Suppose there are three alternative alignments and that all voters agree on the safety/costs scale as the one best catching the salient issue. I presuppose that construction costs and traffic safety are effects with appeal to so different values that voters are unlikely to prefer both the safe alternative and the cheap alternative to the compromise. When this is true, the compromise alternative is ranked third by no one, and the preferences are single-peaked. Thus, single-peakedness permits the voters to disagree sharply on which alternative is wanted, but they have to interpret the problem in similar ways. It is an important practical point that there does not have to be a complete consensus even on the relevant issue dimension. The probabilities of the occurrence of cycles including the alternatives preferred by majorities, reduce to 0.02 and 0.04, respectively, when as few as three-fourths of 45 and 95 voters agree on the issue dimension (Riker, 1982, p. 128).

This point is worth stressing: single-peakedness implies that voters place the alternatives in the same succession along the common scale. In my example, they must agree on which alignment is the safest, which is the compromise, and which is the cheapest in terms of construction costs. As far as the project in question is concerned, "they agree entirely on how the political spectrum is arranged. This kind of agreement is precisely what is lacking in a cycle, where voters disagree not only about the merits of alternatives but even about where alternatives are on the political dimension" (Riker, 1982, p. 128).

When applying social choice theory to transportation or urban and regional planning, for instance, the choice set does not consist of, for example, electoral candidates, but rather alternative supplies of public goods. In economic modelling of consumers' choice among economic options, some restrictions are always placed on their preference orderings to ensure well-behaved utility functions. This diminishes the importance of Arrow's theorem for economics, as his (U)-condition is not observed in that discipline. Much work has been done recently to investigate if a rational and democratic social choice procedure is possible in economic environments (Donaldson and Weymark, 1988). In such environments the choice set consists of public or private goods or some combination thereof. According to Redekop (1995), Arrow-like inconsistencies arise in the interesting economic environments that have been studied.

Consider now condition (I), the independence of irrelevant alternatives. Such alternatives are better thought of as unfeasible or extraneous, as they are merely alternatives that do not happen to be in the set from which the actual choice is made.[6] (I) implies claims for ordinality and for the pairwise-choice format. I explain below what is implicated by these requirements, at least to the extent that it can be judged whether they are violated by synoptic and incremental planning.

Ordinality means that the device for aggregating individual preferences is not to respond to any information concerning the preferences of any individual towards any alternative except that individual's ranking of the alternatives. Preference intensity (cardinality) is thereby ruled out. It is also assumed that information justifying interpersonal comparison of utility is not available, which is the predominant opinion—at least in theoretical welfare economics (Arrow, 1963, pp. 109–118). A measure of the utility has no empirical meaning, because any course of behaviour that can be explained by a given utility function, can be equally well explained by any other utility function which is a strictly increasing function of the first. When a certain level of utility cannot be derived from observed behaviour, we cannot have interpersonal comparability of utilities (Arrow, 1963, p. 9). The behaviour of different individuals can be compared, but since the behaviour does not reveal a specific level of utility, their utility levels are not comparable.[7]

Arrow bases his insistence on ordinality on "the application of Leibniz's principle of the identity of indiscernibles. Only observable differences can be used as a basis for explanation" (1963, p. 109). This implies (a) orderings and not intensities, (b) orderings over a relevant set only, as "there would be no way to observe preferences among alternatives not feasible for society" (1963, p. 110). Strategic behaviour, deliberately misrepresenting preferences for infeasible alternatives, could influence the social choice if (I) is not observed. Since no one else can see my true ranking, my dishonest reporting of preferences cannot be prevented (Plott, 1976, pp. 534–543).

Independence also requires that the way alternatives of any subset are ranked in the social ranking of the whole choice set be independent of information concerning alternatives not in that subset. A pair being a subset, the relative positions of any pair in the social ranking of any number of alternatives must depend only on information concerning that pair (MacKay, 1980, p. 79). The aggregation device wanted by Arrow cannot respond to information about other alternatives than the pair in question, nor to any other information not about this pair. For example, universal generalizations about all alternatives in a choice set cannot be utilized (MacKay, 1980, p. 97). The pairwise-choice format 'turns out to be a way of guaranteeing noncreativity on an ordinal base' (1980, p. 92).[8]

Noncreativity means that we do not want the aggregation device to arbitrarily create the data to which it then responds. An example is the simple heuristic evaluation technique following this procedure:[9]

—Identify a set of consequences of the plan.
—Rank the planning alternatives no. 1, no. 2, etc. for each consequence.
—Add the rank-numbers for each alternative.
—Choose the alternative having the lowest sum of rank-numbers.

This rank-order technique is supplied only ordinal information, but is ruled out by the demands of the pairwise-choice format. In order to determine how many points to give road investment alternative X on (e.g.) the noise reduction indicator, one has to know the noise attributes of *all* the other road investment alternatives. The point I want to make, however, is that the 'distances' between rank-numbers are interpreted as crude measures of differences in quality, and "the technique itself imposes the numbers it then proceeds to respond to" (MacKay, 1980, p. 84). Whatever sorts of data we think it desirable that a device be responsive to, we want the aggregation device to be sensitive just to that and nothing else. The general point is that we do no want social choice to be affected by unintended factors.

When the (I)-condition is accepted, choice rules depending on notions of equality of welfare are not valid. For instance, the maximin criterion of Rawls (1972, pp. 152–153) states that one should choose the alternative which maximizes the welfare of the worst-off person. However, condition (I) requires that the social ordering of a given set of alternatives depend only on the individuals' preference orderings of those alternatives. Restricted to such information we do not know who is worst off in a certain situation, and we do not know if person X is better off in alternative A than person Y is in alternative B (Sen, 1970, Ch. 9).

In concluding this section I note that Arrow's conditions do not imply that preference intensities, income distribution, justice and fairness, truth, etc. are irrelevant to the social outcome. Such input is often crucial when each chooser establishes his or her individual preference ranking. But the decision device amalgamating these rankings is not allowed to take such input into account. Politics influencing the individual preference rankings, forming coalitions, controlling the agenda, manipulating the choice set, etc. is exogenous to Arrow's model formulation. Planners do much of the same things, although based on the authority of expertise instead of the formal authority of the politicians. This means that public planning is also exogenous to Arrow's model. The concept of rationality in social choice reflects this state of affairs. In Arrow's theorem rationality means only connectivity and transitivity, and so it cannot express moral values. In politics and planning it is essential that individuals can be rational not simply in terms of how consistently or optimally they order alternatives, but in terms of the reasons that can be advanced to justify their preferences. However, Frohock (1987, Ch. 5) explains that reasoned forms of rationality do not go well with arithmetical composition rules.

2.2. EXOGENOUS VARIABLES IN PLANNING AND SOCIAL CHOICE

Arrow's formulation of the social choice problem is based on some implicit assumptions in addition to the four explicit conditions presented in the preceding section. One additional assumption is transitivity, as already mentioned. Here, I focus on the variables considered exogenous in Arrow's model, viz., the individual preference orderings, the alternatives in the choice set, and the set of individuals whose preferences are to count (Hylland, 1987). The question is whether it is reasonable to take these variables as given in a public planning process. Will a negative answer cast doubt on the relevance of Arrow's result to a discussion of the possibility of arranging satisfactory planning processes?

A complete planning process includes an evaluation phase laying the foundation for a recommendation. Sometimes the evaluation is guided by a formal evaluation technique. Some of these, like the cost–benefit analysis, produce a list ranking or rating the alternatives. Other techniques merely provide a systematic basis for comparison and leave out the final ranking—like the planning balance sheet (Lichfield, 1995). A formal evaluation technique ranking the alternatives is the clearest analogy in the planning process to the aggregation device searched for in Arrow's problem. Yet plan-making is so much more than evaluation, as is obvious from the phases of problem identification, goal formulation, option seeking, and impact assessment. The efforts to sort out the problem and process the information to illuminate it may influence factors that Arrow considered exogenous.

I will first comment on the set of people whose preferences are to be taken into account—'the choosers' for short. It is usually clear who are members of the political body or assembly formally making the decision. But in an indirect democracy they often represent others. Some of the represented may be particularly affected by a planned action and may thus be granted special opportunities for influencing the representatives. Who is directly and strongly affected will sometimes be uncertain until the impact analysis has been carried out. If so, the set of choosers may not be exogenous; it is not necessarily known prior to the planning process.

When it comes to the planning alternatives, it is plain to see that the choice set is often not exogenously given. The option seeking phase implies just that. In principle, the choice set will not be known until the impacts have been analyzed. The endogenous nature of the choice set is underlined when metapreferences are accepted in the analysis (Grofman and Uhlaner, 1985). Metapreferences concern the choice rule itself or characteristics of the decision-making process. Thus, each alternative consists of the combination of process and product, and it does not seem reasonable that all relevant processes are given in advance by others than the planners. Besides, the outcome of the decision-making process can be influenced by those in control of procedures manipulating the agenda (Riker, 1982, Ch. 7). As long as manipulation is possible by restricting alternatives or introducing a suitable number of new ones in order to change the rank order of those already known, the choice set is endogenous.

Then there are the individual preference orderings. Preferences change as new opportunities are identified, as new information on existing opportunities becomes available, as persuasion is successful, as questions are asked and attention drawn to new issues. In practical planning there is a continuous pressure to alter the preferences of those involved. The formation of preferences is often one of the main reasons for going through the planning process at all, hence the preference orderings are endogenous.

If the people involved are fully communicatively rational in Habermas's sense and always manage to make consent the outcome of their interaction, there is no room for any aggregation device like Arrow's. The problem of social choice would then be solved, but this is very far from reality.

One will see that the reasonableness of regarding the choice set, the preferences, and the set of choosers as exogenous depends on the planning theory that frames our ideas. The simplest form of the synoptic planning process is linear, and when the evaluation phase is reached, all necessary information has been processed. Then the acceptance of Arrow's presupposition of what is exogenous is not problematic.

The learning and loops of the incremental process blur the distinctions between the planning phases. Moreover, there is no consideration of the optimal search for information; the first satisfactory solution is implemented. However, the solution may be counteracted by ensuing decisions. Both the preparation for making a particular decision and the sequence of decisions can be looped. Furthermore, posterior reasoning may continue to change preferences throughout. In a purely incremental process one may never reach the point beyond which the set of alternatives and individual orderings can be taken as given. The differences between the planning modes as seen from the perspective of social choice theory are scrutinized in the next two chapters.

We have seen that planning sets the amalgamation of individual preferences into a broader context in which there are more endogenous variables than in Arrow's formulation of the aggregation problem. What is essential here is that the planning context is not, in general, likely to facilitate the combination of democracy and logicality. On the contrary, by deliberately changing the set of choosers, the alternatives on the agenda, or the individual rankings, the option first chosen may be downgraded to a loser. In fact, working for such changes is an integral part of politics. Thus, if choosers, agenda, and individual orderings are considered as endogenous variables, Arrow's theorem is even more of a troublesome reality. However, it will be seen in Section 5.2 that dialogical features of the planning process may pull towards the opposite conclusion.

2.3. THE PLANNING PROCESS AS AN ARROW TYPE AGGREGATION DEVICE

The assumption of path independence indicates that there is some inconsistency between planning theory and part of social choice theory. If the outcome of the social decision procedure is independent of the path leading up to the final choice, then public planning is without significance. However, path independence is usually defined more narrowly as independence of the sequence of voting. Planning can be regarded in two distinct ways from a social choice point of view, and the first is—to a higher degree than the second—in line with the distinction between planning and decision-making implied by path independence.

(1) Planning is seen only as preparation for decision, and decision-making is viewed as separable from planning. The question is then if planning can strengthen the interpretation of the collective decision as the will of the majority. Planning can sometimes make single-peaked performances more likely by providing information stimulating a degree of conformity of preferences and by simplifying the options.
(2) Public planning is seen as an integrated part of the political decision process. In practice much is decided outside the City Council, both before and after the formal political decision is made. Planning techniques and even the entire planning process may hence be studied as a social choice process.

Most of the essay examines planning in the second perspective. Social choice theory tells us that no decision device satisfies all the claims that seem reasonable when made separately. The question is then what claims or values the modes and techniques of planning compromise.

First, however, some remarks based on the first angle are presented below. I spell out briefly what the most well-known modes of planning do to promote single-peakedness. The first perspective is picked up again when discussing dialogue and social choice in Section 5.2.

The two approaches above are to some extent combined in Chapters 3 and 4 on synoptic and incremental planning. It is interesting to investigate, for example, if incrementalism pushes the preferences of the choosers towards single-peakedness and also relaxes the (U)-condition in a way that would not be so objectionable were the individual preferences very likely to be single-peaked.

An important task of planning is to cater for the informational needs of the decision-makers. In perspective (1) above, planning complements politics in its formation of the individual rankings that are eventually 'handed over' to the formal decision device. Planning also provides information which can be used politically to determine the set of alternatives and thus the agenda. The various planning modes constitute different strategies for making an impact on the decision-makers' rankings, and they stress efficiency and distribution to different degrees. Beneath I comment on those strategies from a social choice point of view.

Synoptic planning relies heavily on professional analysis. The planners collect enough information to decide what solution is the optimal one under specified assumptions. The criteria employed when deriving the optimal solution are rarely sufficiently comprehensive to bring all informed opposition to silence. Suppose, nevertheless, that the decision criteria reflect the values of a great majority. Then the extensive impact assessment of synoptic planning is often an effective tool for making individual orderings converge, resulting in value restrictions sufficient for single-peaked preferences.

Incrementalism will probably affect the choice set to a greater extent and individual preferences to a lesser extent compared to synoptic planning (Lindblom, 1959). In disjointed incrementalism one aims to identify solutions which most of the decision-makers agree on. Big alterations affect the community in many ways—social, political, economic, and environmental. Hence, people are likely to assess radical change from several angles. The possibility of designing projects that affect the community mainly on a single dimension increases as the scale is reduced. It is usually sufficient for avoiding Arrow's paradox that most choosers agree on the dimension; unanimity is not required. Disjointed incrementalism will thus often be able to prevent Condorcet's paradox in each particular vote. Because of the lack of goals providing direction, cycles may nevertheless occur over time.

Transactive planning is not focused on processed knowledge and analytic means–end models but places more emphasis on the deployment of the personal knowledge of local citizens (Friedmann, 1973). It is a main idea that the choosers develop insight into each others' arguments and know and understand the reactions of those affected by the plan. The stronger demands on the quality and openness of communication, the more limited is the scope for egoistic argumentation. When the preference rankings of the individual choosers can be scrutinized by the public, the orderings are likely to be affected by this fact. The effect is reinforced by the development of close and personal relations intended to stimulate mutual empathy concerning problems and aspirations. It can be assumed that this planning mode induces a degree of conformity on the way to conceive problems and the criteria for assessing planning alternatives. Then the voting paradox is less apt to arise.

Advocacy planning conveys selective knowledge to the choosers (Piven, 1970; Heskin, 1980). They are informed that the distribution of costs and benefits is biased. Or they get to know that those disadvantaged by the plan are, roughly speaking, those that are already getting less than their fair share. By pointing out one or more of the planning options as particularly unfair, advocacy planning contributes to the formation of value restrictions guarding against the perverseness of the social choice paradox.

Critical pragmatism (Forester, 1989, 1993; Sager, 1994) generalizes the advocacy model. It is based on a critical attitude towards authorities and biased power relations rather than commitment to a particular group. Critical pragmatists counteract systematic sources of misinformation. They help citizens reveal attempts at misrepresenting cases, improperly invoking authority, making false promises, or distracting attention from key issues.

Forester (1993, p. 27) contends that:

In the communicative dialogue, conversation, and play of power that constitute the planning process, the evolution of the questioning of selective possibilities and the shaping of equally selective responses, planners' communicative actions work to organize (or disorganize) citizens' attention, their engagement, investment, and participation...(P)lanners open or foreclose possibilities, alert or ignore others, call forth or disregard particular concerns, and spread or narrow the bases of design, criticism, participation, and thus decision-making.

The planner is not seen as a processor of facts, but as a practical organizer of attention. Regarded as organizing, planning implies advancing some views and issues and holding back others. A central admission of John Forester's critical theory of planning is that communicative rationality is always imperfect. In his view, the ethical framework for the everyday practice of planning analysts is to eliminate distortions, to foster open and authentic communication, and to make true political discourse and dialogue possible. This is the paradigmatic core of 'planning as questioning and shaping attention', which is a frequently used characterization of critical pragmatism.

These keywords indicate how this mode of planning helps us to circumvent Arrow's paradox. Firstly, being questioned by planners, local people and various interest groups, the choosers will feel a pressure to back their preference rankings by arguments suited for public debate. The effect of this is already described under transactive planning above. Secondly, by shaping attention, the planners influence the choice set. It was described under incrementalism above how this can be used to increase the likelihood for single-peakedness.

NOTES

1. Transitivity means that if alternative X is preferred (indifferent) to alternative Y, and Y in turn is preferred (indifferent) to Z, then X must be preferred (indifferent) to Z. Connectedness means that for any two alternatives X and Y, either X is preferred to Y, or Y to X, or they are indifferent (MacKay, 1980, p. 7). Intransitive social ordering implies that cycles can occur, i.e. one alternative is ahead of a second, the second is ahead of a third and so on until the last is ahead of the first.

   A Condorcet winner beats or ties all other alternatives in paired comparisons. When there is no Condorcet winner cycles occur and the background condition of transitivity is broken.

   Note that several other properties seem reasonable to demand from collective choice functions; for instance, anonymity, neutrality, and responsiveness (Sen, 1970, p. 68). 'The conditions that Arrow showed to be inconsistent are not conditions that he regarded as *sufficient* for a satisfactory system of collective choice, but what appeared to him to be plausible necessary conditions' (1970, p. 49). As Sen goes on to say, a social choice function "can pass all the tests of Arrow and still look very unappetizing" (1970, p. 50).

   It has been proved that impossibility theorems similar to that of Arrow can be constructed by requiring combinations of other conditions than those suggested by Arrow, for instance the impossibility of a Paretian liberal (Sen, 1970, Ch. 6; Sen, 1983; Aldrich, 1977). However, I will delimit my discussion to the four conditions of the original and celebrated theorem.
2. The general impossibility theorem was originally stated in terms of orderings. Arrow looked for an acceptable social ordering of all the alternatives. Later stu-

dents of social choice objected that a complete ordering is not important compared to the collective ability to always pick *one* preferred alternative. The problem can be restated in terms of choice rather than ordering, and the main defect of the method of amalgamation is unaffected by the new language (Riker, 1982, p. 133).

3. Riker cites the following statement by Plott: "(T)he process of choosing, from a dynamic point of view, frequently proceeds in a type of 'divide and conquer' manner. The alternatives are 'split up' into smaller sets, a choice is made over each of these sets, the chosen elements are collected, and then a choice is made from them. Path independence, in this case, would mean that the final result would be independent of the way the alternatives were initially divided up for consideration." (Riker, 1982, p.134).
4. An oligarchy is a subset of choosers who, if the members agree, can impose a choice, or, if they do not agree, enables all members individually to veto the choice (Riker, 1982, p. 131).

   The question of accepting a dictator is sometimes considered when commenting on Arrow's theorem in light of John Rawls's (1972) theory of justice (Strasnick, 1976). Putting heavy emphasis on fairness, one can find a way out of Arrow's paradox by accepting as a dictator the one who has more to lose in the collective outcome. Tanguiane (1994) searches for a solution to the social choice problem in a refinement of the definition of a dictator. He proves that there always exists an individual who, on average, represents a majority preference, and he limits the concept of a dictator to the dictators who are not representatives. Then he proves that a non-dictatorial social choice, satisfying Arrow's axioms, is possible.
5. Sen (1985, p. 1770, Note 19) asserts that: "In fact, the condition can be a little less exacting, and applied to each triple separately. The condition is equivalent to a unanimous agreement that in any triple (x, y, z), one state (say, x) is 'not worst'. The condition can be readily extended to unanimous agreement on some state being 'not best', or unanimous agreement that there is a 'not medium' state. This more general condition is called 'value restricted preferences'."
6. Schwartz (1986, Ch. 10) gives attention to how one should be choosing the set from which to choose. This involves the awkward problem of how to determine what is feasible. The reason for drawing a line between feasible and infeasible is to limit the scope of our reasoning about the choice in question. "Unable to reason about everything at once, humans must take some things for granted" (1986, p. 227).
7. Seabright (1989, p. 371) reminds us to distinguish cardinality and comparability: "Extending the utility information to include cardinal utilities does not avoid Arrow's result unless interpersonal comparisons are also permitted (utilitarianism makes use of both). Extending it to allow interpersonal comparability without cardinality does help, however. In fact, a version of Rawls's maximin criterion is consistent with the (extended) Arrow conditions; it requires that we identify the worst-off individual in each state and compare the welfare of the worst-off individual in one state with that of the worst-off individual in another, but strengths of preference need not be taken into account."

Rawls's maximin criterion is not the most obvious in planning modes firmly grounded in instrumental rationality (Sager, 1993). It is, however, quite compatible with advocacy planning.

8. MacKay (1980, pp. 92–101) tries to distinguish noncreativity itself from the means used to achieve it, namely, pairwise choice. He strongly suggests that giving up the pairwise-choice format is a plausible way to remedy against Arrow's paradox.
9. This procedure is known as the Borda count. It is axiomatically treated by Young (1974), and Plott (1976, pp. 516–517, 560–563) gives an informative account in more ordinary language. Plott contends that the Borda count does not violate the (I)-condition as formulated by Arrow (Plott, 1976, pp. 535–538). Riker (1992) draws the opposite conclusion, and I follow him.

CHAPTER 3

Synoptic planning as relaxation of the Arrow conditions

The task is now to examine whether synoptic planning comprises procedures or theoretical elements that can help to evade the paradox of social choice. The chapter spells out in what ways synoptic planning relaxes Arrow's (I)-condition and leads to a compromise between the qualities of logicality and fairness by yielding on other aspects of rationality than transitivity. The first section takes a brief look at the various phases of a synoptic planning process and inquires if the task set for each stage has implications for any of the Arrow's conditions. The second section deals with the evaluation phase. An impossibility theorem is identified, which preserves the logical structure of Arrow's theorem. The correspondence between formal evaluation techniques and the presuppositions of this new impact assessment theorem is examined. The third section explains how logicality is modified, especially by relaxing the pair-wise-choice format and permitting information in addition to the individual preference orderings.

As an 'ideal type', synoptic planning might be specified as being without limitations on obtainable information. In that case, the intensities of one's own and others' preferences are known to each individual and are fully comparable. Then Arrow's social choice problem does not arise. Numbers can be attached to the utility levels, and the planning alternative resulting in the greatest sum of individual utilities can simply be calculated. Not only utilitarianism, but even several other decision rules satisfy the Arrow conditions when the poverty of information is eliminated. Even if there is still the ethical problem of equity vs maximum collective utility, social choice becomes really intriguing only when the perfect information assumption is modified.

I assume here that information in synoptic planning is less than perfect, in that preference intensities and interpersonal utility differences are not known. Furthermore, computation takes time but is otherwise supposed to be free. The calculative capacity is sufficient to identify the best alternative and to survey the effects of any action under consideration. When all information necessary to carry out a certain task is either available already or can be calculated while solving the task, there is nothing to gain from returning to that task later on. The assumption of intersubjective knowledge opens for relaxation of this linearity condition. The planners may now increase their knowledge as time goes by; that is, the number of issues agreed upon may expand throughout the process. Then one can imagine situations in which the improved insight is important enough to warrant reconsideration of previous work.[1]

3.1. SOCIAL CHOICE IMPLICATIONS OF THE SYNOPTIC PLANNING PHASES

Most main features of synoptic planning do not imply violation of Arrow's conditions. Abundant information and certainty about which solutions serve each individual best, provide a complete picture of causal relations and eliminate misunderstanding and unrealistic expectations. Indeed, this can rule out some problem formulations and some proposed solutions. But I do not assume that consensus is a necessary or even likely product of knowledge. In a synoptic context one can perceive of a problem in different ways, and there is no guarantee for single-peakedness. When this planning mode modifies the Arrow conditions, it is because of the intrinsic maximization and the analytic techniques applied to achieve it; above all, the evaluation techniques. I will examine if the phases of the synoptic process are of any significance to the possible relaxation of Arrow's (U) and (I)-conditions.

Initially I make two general observations. First, (I) requires that neither global information nor cardinal information is taken into account. Second, synoptic planning is an optimizing method and goes for the 'best' solution. The question is whether the two claims are compatible. Is it possible to maximize without using global information? Such information is about the whole choice set and does not refer to a less-than-total subgroup only. Whether a given alternative (X) is a maximum alternative in a setting is a global fact about that setting. We not only need to know whether X beats B, C, D, etc. in pairwise comparisons. We also have to know that all alternatives have been checked, and that the specified alternatives do in fact exhaust the feasible set. We thereby utilize a fact about how one alternative relates to all the rest (MacKay, 1980, p. 94). The feasible set is exogenous in Arrow's problem formulation, so when starting the maximization the necessary global knowledge is already at hand. A possible restatement of (I) would be that except for the global information implicit in the exogenously given feasible set of alternatives, only information on the individual orderings of the alternatives is fed into the social amalgamation device.

Maximization requires that the goal-variables are commensurable in the sense that one can trade an increase along one goal achievement scale with a decrease along another. Figure 1 depicts a situation in which both the construction cost/traffic safety dimension and the time cost/traffic noise dimension are judged to be relevant but do not have to be considered separately. A and B are planning alternatives, and the figure shows how the choosers situate them on the issue scales. If the choosers are to agree on the locational ordering of the planning alternatives along the more general economy/quality-of-life issue dimension, they must be able to trade off gains in safety and noise. This commensurability required by the synoptic approach paves the way for single-peaked preferences.

I now turn to some of the phases of a synoptic process, and evaluation will be given most attention. First, however, some remarks on the early phases of the process: the problem identification does not in any way guarantee that a single issue

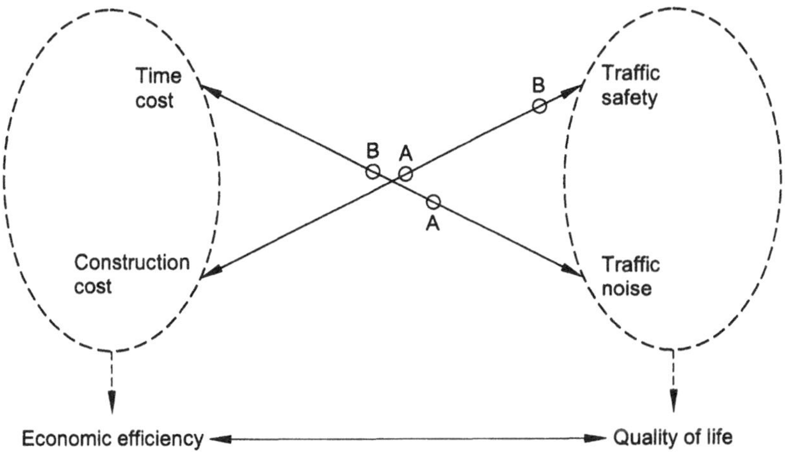

FIG. 1. Reducing the number of issue dimensions.

dimension can be found and agreed upon. But the important dimensions will be discovered and the main trade-offs will become clear. This is a necessary structuring before one can proceed to investigate whether there is a single and more general issue dimension preserving what is essential to the majority in its understanding of the problem. Formulation and analysis of goals and objectives often overlap the problem identification. If we imagine that a hierarchy of goals is worked out, the possibility for agreement on a single issue dimension is highest when the top goal can be divided into just two goals at the next highest level. These may then constitute the end points of the issue scale. In Fig. 1 one can imagine economic efficiency and quality of life as resulting from the splitting of a general welfare goal.

Option seeking eliminates infeasible alternatives and sees to it that all the feasible ones are in the choice set from early in the decision process. Then it is not possible to manipulate the voting result by bringing in new alternatives on suitable occasions.

Means–end analysis and prediction cast in the form of impact assessment check if the alternatives generated do in fact deserve to be in the final choice set. The broad search for alternatives creates a need for an analytic technique helping to reduce the set of proposals before the final consideration by the decision-makers. Impact assessments always contain information measured along the interval and ratio scales. Each individual can use such information when balancing the effects on (for example) noise and traffic safety to work out a personal preference ordering. But the impact assessment results cannot be brought directly to bear on social choice and supplement the information base consisting of the individual rankings.

In the evaluation phase of synoptic planning, analytic technique is frequently employed. Most evaluation techniques aim to reach a recommendation on which option the collective ought to choose. Provided that the conditions (U), (P), and (D) are to be observed, a recommendation cannot be deduced merely by permitting intensities of utility (cardinalism).[2] It must in addition be considered possible to compare the utility scales of different individuals. The first assumption made in economic evaluation technique (the CBA) to make such a comparison seem plausible, states that every variable to which preferences are attached, can be traded off in monetary units. The problem is then limited to the question of whether the individuals get equal utility from a certain amount of money. The biased distribution of income is a prominent reason why the marginal utility of money varies among individuals. The second assumption of the CBA made to sustain the comparison of utility scales is that a benefit of a certain amount is equally valuable to society no matter to whom it accrues. That is, the technique ignores the adverse consequences of a biased income distribution on the reasonableness of interpersonal comparison of the rankings of the planning options. In general, when the (I)-condition is ignored and cardinal information and interpersonal comparison of utility are accepted, a utilitarian calculus like the CBA satisfies the other Arrow conditions and provides transitive collective orderings (Hylland, 1987, p. 20).

Let us now assume that an aggregation device is wanted which fulfils Arrow's conditions, and that the rank order of each individual in the electorate is to count. However, we also presuppose that the decision among the alternatives will actually be made by representatives, say by a county council. Then any use made of the results of an evaluation technique by the council violates the condition (I). In fact, any special attention given to the conclusions of policy analysis in collective decision-making produces this result, as it "involves going beyond citizens' preferences and involves criteria derived from analysis of the problem itself" (Douglas, 1984, p. 75).

A strong and 'closed' planning agency in a public sector, e.g. the Public Roads Administration, may reinforce the tendency to relax condition (I) by means of analytic technique. It can be less open to input from pressure groups and participating citizens. What will most typically be conveyed to the politicians is not the outcome of communication with the public, but the result of analysis. Analytic problem-solving may replace political bargaining as the bureaucratic ideal.

Assume now that a group-based formal evaluation technique is employed. This technique leads to a set of rankings of the planning alternatives seen from the perspectives of different affected groups. The remaining step, i.e. the final aggregation over groups, is a close parallel to Arrow's problem. The only difference is that one is to amalgamate group orderings instead of individual orderings. At this stage, now that group orderings are already established, it might be asked why the procedure used to derive the group orderings cannot be applied once more to produce the social ranking. This is not possible, however, without a fundamental departure from the Arrow conditions. The reason is that the evaluation technique does not receive the preference orderings of each unit to be aggregated (groups in this case) as input, and so it breaks condition (I). The aggregate ranking calculated by an evaluation technique is thus not equivalent to the social preference ordering Arrow was looking for. In fact, cost–benefit analysis (CBA) and similar techniques are used partly because we do not find what he searched for.

3.2. THE IMPACT ASSESSMENT PARADOX

The previous section showed that synoptic planning relaxes the Arrow conditions mainly by employing policy analysis, i.e. impact assessment and evaluation technique contradicting (I). A closer consideration of the evaluation method is therefore justified. When the evaluation phase is reached, the results of the impact analysis may be deployed in a formal evaluation technique, like the goals achievement matrix or the cost–benefit analysis. One can reformulate Arrow's four conditions to make them explicitly apply to analytic evaluation and aggregation of different planning impacts. This is done below in a way preserving the logical structure of the original impossibility theorem. The reason for the exercise is to examine if the transformed theorem can shed light on how the evaluation phase of synoptic planning tends to relax Arrow's conditions. The transformed set of conditions concerns the aggregation of rank orders over all the alternatives for each consequence of the plan, not the aggregation of rank orders over all the alternatives for each individual chooser. Hence, the single consequence here takes on the role that the single individual has in Arrow's problem formulation. This immediately suggests that the adverse effect of relaxing some of the conditions below may be less serious than in Arrow's case, since consequences do not manipulate like individuals do.

(U_0): Any number of consequences and planning alternatives is allowed, and the alternatives can be ranked in any order for any consequence.

(P_0): If one alternative gives higher goal achievement than another for every consequence, it defeats the other in the overall ranking—i.e. domination is respected.

(D_0): The overall ranking is not to be dictatorial. The aggregation device is said to be dictatorial if there exists a consequence so that the overall ranking of the alternatives is always the same as the ranking with respect to this consequence.

(I_0): The overall ranking of a given set of planning alternatives depends only on the orderings of those alternatives with respect to each of their listed consequences.

In Arrow's approach the preference ordering of each individual should be taken account of by the aggregation device. In the problem at hand this corresponds to the evaluation technique taking account of 'each consequence's ranking' of the alternatives. (U_0) implies that the specified consequences are independent in the sense that a particular ranking for one consequence does not logically restrict the way alternatives can be ranked on other consequences.

Since the logical structure of the problem is preserved, the paradoxical result is retained.[3] What we can denote 'the impact assessment paradox' says that no evaluation technique can simultaneously fulfil the four conditions (U_0), (P_0), (D_0), and (I_0), and guarantee a transitive ranking of the planning alternatives. This, of course, only affects techniques leading all the way to a ranking of the alternatives; hence, not the planning balance sheet and techniques relying on nominal scales only.

Some techniques are lexicographic, and this property can refer to goals or—as it does here—to consequences. Just one 'consequence's assessment' of the alternatives is to count. Only when alternatives tie over the most important effect is the ranking for the second-most important consequence to count. In the Arrowian scheme, this is equal to the contention that only when the dictator is indifferent are the preferences of others to count. This kind of lexicography thus violates the condition (D_0). Lexicography is similar to incommensurability in that changes of different consequences cannot be traded off. Any small improvement in the most important type of impact for an alternative has a greater influence on the overall ranking than whatever vast change for the better or worse that might take place in other impacts. Incommensurability may be taken to mean that consequences cannot be compared at all, i.e. that they cannot be amalgamated. Then an evaluation technique ranking the alternatives on the basis of the rankings for each particular consequence cannot be constructed.[4] Lexicographic ordering is not contrary to (I_0), but the ranking provided is not really overall.

As far as I can see, except for lexicographic techniques, all well-known evaluation techniques in planning satisfy conditions (U_0), (P_0), and (D_0). However, all the well-known techniques offering an overall ranking of the planning alternatives violate (I_0). The simplest of the rank order techniques violates the pairwise-choice format and thus the independence part of (I_0). The technique consists of ranking the alternatives for each consequence, then counting the number of top-rankings for each alternative. The alternative with the highest number of top-rankings is chosen. Only when the two best alternatives have the same number of top-rankings is the number of consequences for which they are ranked second of any interest. If an alternative drops out of the choice set, it can influence which of the remaining alternatives obtains most top-rankings. An alternative, originally scoring second on several of the consequences for which the withdrawn alternative earned top-rankings, will be given an advantage. When there is no tie, this simple evaluation technique works as an analogue to the ballot, choosing the candidate receiving the highest number of votes. The ranking procedure requires information about all the alternatives.

I have already discussed the evaluation technique recommending the alternative having the lowest sum of rank numbers when adding over all the consequences. This technique (the Borda count) violates the pairwise-choice part of condition (I_0). All evaluation techniques applying interval scales, such as point rating techniques like the goals achievement matrix, or ratio scales, such as the cost–benefit analysis, violate the ordinality part of condition (I_0). When (I) is violated one employs information that is unavailable in principle, while (I_0) can be violated by applying readily available knowledge—as will be shown shortly.

As nearly all the evaluation techniques violate the condition (I_0), it could be argued that more rewarding and discriminating results might be reached by relaxing (I_0) and making the other conditions more rigorous. While this may be so, the conditions just considered indicate that a standard way for planners to by-pass the impact assessment variant of the paradox of social choice is to relax the equivalent to Arrow's (I)-condition. In the Arrow context, preference intensities may be ruled out because they are considered unreliable, or because one does not want to give fanatics more influence than moderates. In the case of the impact assessment paradox the 'ordinality' condition would be relaxed by taking account of directly measurable differences between the alternatives with regard to a particular consequence, e.g. noise or maintenance cost. The 'fanatics argument' has no bearing on this case, and the required information is often taken to be objective, uncontroversial, and easy to collect. I conclude that the 'ordinality' part of condition (I_0) may well be too restrictive.

It is of interest to social choice how evaluation techniques relate to the conditions (U_0), (P_0), (D_0), and (I_0), even if choice is determined by the summation of the individual votes of elected decision-makers. The outcome of policy analysis, like the cost–benefit analysis (CBA), can influence and even replace the rank orderings of decision-makers.[5] I distinguish three cases below.

In the first and rather extreme case, the decision-makers hand over their authority to the technocrats (planners) by accepting the result of the CBA whatever their personal rankings of the planning alternatives may be. This situation might be the result of severe internal conflict in the decision-making body. The effect is that the condition (D) is broken, because the technocrat (planner) functions as a dictator to the extent that his or her interpretation of the public interest affects the result of the CBA.

In the second case, the recommendation of the CBA is not necessarily approved by the decision-makers. However, the analytic technique has shaped their view of reality so that a single issue dimension emerges, formed by concern for the economic welfare of society at the one end of the scale and by consideration for the environmental and social intangibles at the other end. The condition (U) is relaxed by technocratically-induced issue standardization: the techniques we apply affect our interpretation of reality.

In the third case, the result of the CBA is taken as just another piece of information about the merits of the alternatives, along with several other items. It influences the preference orderings of the decision-makers on a par with the knowledge of other properties. Then the characteristics of the CBA are of no particular interest in connection with the Arrow paradox.

I conclude by giving a stepwise recapitulation of the logic of presenting the impact assessment paradox, adding a couple of points which explicate the difficulties it creates for planning:

(1) The evaluation phase is the one most closely affiliated with Arrow's problem, and a formalized evaluation technique provides the nearest analogy to a device for social rank ordering.
(2) Elaborate evaluation techniques based on interval or ratio scales (e.g. the CBA) are designed for synoptic planning rather than assessment of incremental, reversible steps.
(3) The Arrow conditions can be reformulated to fit the problem of plan evaluation, viz. to let information on each of the impacts count in a formula providing a total ranking of the alternatives.
(4) The transformation of the conditions preserves the logic of Arrow's problem and generates a related impossibility theorem.
(5) This impact assessment paradox does not cause much trouble, however, because one of its conditions can be relaxed without serious consequences.
(6) The parallel to breaking the ordinality condition and accepting preference intensities in Arrow's problem is, in the case at hand, the calculation of differences between alternatives for the consequences of the plan. When the differences are measured for a goal variable, information about goal achievement is obtained.

For example, the goal may be a specified reduction in traffic accidents, and we can calculate the expected number of such accidents for alternative plans.

(7) Impacts are often quantified so that the differences between the alternatives can be calculated without resort to preference intensities. Then evaluation in synoptic planning can violate (I_0) and by-pass the impact assessment paradox in a way that does not in itself break the Arrow conditions. Relaxing the 'ordinality' part of (I_0) does not necessarily violate the ordinality part of (I).

(8) However, there are impacts for which quality differences cannot be measured in an objective way with the knowledge currently available. These should be excluded from the evaluation technique when one wishes to observe the Arrow conditions, because subjectivity and the display of preference intensities are easily mixed up.

This discussion of the impact assessment paradox indicates that the evaluation of planning alternatives founded on 'fair' and transitive aggregation of information on the single impacts, is in principle possible without violating the Arrow conditions. In this conclusion, 'fair' is as seen from 'the viewpoint of consequences' instead of individuals.

3.3. SYNOPTIC PROCEDURES WEAKENING LOGICALITY

Fairness and logicality cannot be simultaneously obtained. Softening Arrow's conditions (U), (P), or (D) would imply lower ambitions concerning fairness and democracy. To lower the ambitions on logicality one has to relax the (I)-condition or modify the transitivity assumption. I will now discuss how logicality is debilitated by synoptic procedures relaxing (I).

Under synoptic conditions as specified in this essay, trustworthy knowledge of preference intensities is not available. It would be illogical then to violate the ordinality part of (I). That would simply mean seeking support in nonexistent information. This criticism affects the application of evaluation techniques based on cardinal information (like the CBA).

The (I)-condition can also be relaxed by violating the pairwise-choice format. In synoptic planning, with ample information on most variables, global information relevant to planning is at hand. If we disregard the demand for pairwise choice, it becomes possible to take into account ordinal information directly from the impact assessment (i.e. information supplementing the individual rankings). Global information from this source can state, for instance, that all the alternatives satisfy the standard set for the maximum noise level. Or it may inform us that none of the alternatives in the choice set conflicts with decisions made in adjoining sectors.

In synoptic planning it seems desirable to make use of global information. It strengthens the basis of the maximizing procedure, and enables us to more fully exploit the rich information from the impact assessment. One can allow for global information by replacing the (I)-condition by an ordinality condition and observing a no longer explicitly stated commitment to noncreativity. MacKay (1980, p. 100) formulates the ordinality condition as follows: "A device shall not respond to any information concerning the preferences of any individual toward any alternatives except information about the order of that individual's preferences". The independence part of (I) is left out and with it the pairwise-choice format. When the application of global information is regarded as an implication of synoptic planning, this planning mode causes modified logicality:

> (I)t will no longer be prohibited for some pair of alternatives to be socially ranked one way in one setting and a different way in another, even though all individual preference orderings remain constant. Everything will depend on the character of the extra factor(s) (if any) involved. The reason for the difference in social ranking will have to be sought out and examined. The difference per se will no longer be decisive. (Mackay, 1980, p. 101)

I take logicality to be qualified in Arrow's approach if social choice is arbitrarily influenced by other factors than individual orderings. But it remains to be seen if fairness can be improved by sacrificing some logicality in this manner. The above examples—observing noise standards and preventing conflict between decisions made in different sectors—indicate that democratic gains are possible from accepting global information, i.e. relaxing the pairwise-choice format. Imagine that decisions are democratically made on all levels and in all areas of public policy. If the overall system is to function well politically, the various democratic assemblies have to respect each others' decisions. Global information can tell us if a choice set has qualities making it likely that this claim is met. There are thus some advantages in dropping the pairwise choice format from a democratic point of view, although there are democratic disadvantages too. "After all, every known electoral system satisfies this condition" (Arrow, 1963, p. 110). One reason why election procedures presuppose (I), is unwillingness to risk that the social ranking of the remaining alternatives is affected if a feasible alternative becomes infeasible and leaves the choice set. Arrow uses the example of a candidate dying while the votes are counted.

The comparison of synoptic planning and incrementalism as strategies for mitigating the serious effects of the social choice paradox will be returned to at the end of the next chapter.

NOTES

1. Even if information is less than perfect, in other ways too, there might be metapreferences in favour of the linearity of the synoptic planning process, because that property reduces decision costs (Grofman and Uhlaner, 1985, p. 42). Concern for decision costs leads voters to care about the simplicity of procedures and to prefer timeliness. The quest for timeliness is manifested, for example, in parliamentary rules which complicate or make impossible reconsideration of previous defeated motions and thereby avoid the spending of time in rehashing 'already decided' issues. Grofman and Uhlaner hold that this and other metapreferences can explain the high degree of stability found in collective decision-making in practice compared to what should be expected from the theoretical results.
2. Cardinalism (knowledge on intensities) and interpersonal comparison of utility levels are different things. Even ordinalism can be combined with interpersonal comparisons. It requires that one is able to decide if person i is better off when x

is realized, than person *j* will be when *y* is implemented. Under these information conditions Rawls's maximin rule is applicable, and it satisfies the (suitably revised) Arrow conditions (Hylland, 1987, p. 20; Seabright, 1989, p. 371).

3. I do not prove that my transformation of Arrow's social choice problem gives paradoxical results. MacKay (1980) discusses a similar transformation and offers an intuitive explanation. He transforms the problem into that of aggregating the results of each single event in a multi-event contest, e.g. decathlon.
4. Incommensurability is rather meaningless the way the main problem is formulated here. It would imply that we try to deduce the collective ranking of units that do not constitute a collective. The analogy in Arrow's problem is that one tries to deduce a social ordering from the preferences of individuals for whom collective decisions cannot be legitimately made, because they are members of different, non-overlapping constituencies. Hill (1968) suggests that goals may be incommensurable. Thus, some versions of the goals-achievement matrix (GAM) are presented as desegregated schemes. See Sager (1981) for an overview of the GAM-variants.
5. Evaluation techniques like the CBA are not likely to be used in incremental planning. In order to clarify the contrast between the planning modes, I nevertheless spell out how disjointed incrementalism relates to the impact assessment paradox.

   Incrementalism violates (U_0). There are restrictions on the number of effects and alternatives that are evaluated. Furthermore, only alternatives relatively similar to the *status quo* are permitted. (D_0) is probably accepted, although the political consequences (agreement) are emphasized. (I_0) is violated by disregarding ordinality, for instance. The most serious, however, is the violation of (P_0). The planning alternatives are compared to the established aspiration level one by one in the same sequence as they are generated. The option first found to pass the test is chosen. Hence, with this strict satisficing, the entire impact assessment paradox is irrelevant.

CHAPTER 4

Incrementalism as relaxation of the Arrow conditions

The chapter completes the analysis of the relationship between the type of collective choice process preferred by society and the synoptic–incremental planning dichotomy. The disjointed incrementalism of Charles Lindblom violates the Arrow conditions in a number of ways, and the purpose of this chapter is to explain which incremental properties have this effect. It will become clear how the relaxations lead to a compromise between logic and fairness in the decision process. Most peculiar to incremental planning are the violations of (U) resulting in modifications of fairness. I separate the features relaxing (U) from those relaxing the (I)-condition. This facilitates the comparison of synoptic and incremental planning as strategies to alleviate the consequences of the social choice paradox.

In a way Lindblom's incremental model seems to make the social choice problem formulated by Arrow irrelevant. In the disjointed and partisan political process envisaged by Lindblom it is hard to imagine any central political body taking the trouble of incorporating all the individual preference orderings into its decisions. Collective institutions responsible for promoting the public interest are not given much attention in his pluralistic and politically competitive society. This is a bit puzzling, as incrementalists seem to need institutions in order to limit cycles and ensure the political stability on which their theory of social decision-making rests. Indeed, the second section will show that there are features of incremental decision-making that conflict with Arrow's formulation of the social choice problem.

4.1. DISJOINTED INCREMENTALISM AS A PROCEDURE RELAXING THE CONDITION OF UNRESTRICTED SCOPE

The section examines how some features of incremental decision-making imply relaxation of Arrow's (U)-condition. The features are (1) neglecting some values and consequences, (2) fragmentation limiting the number of people making each decision, (3) moving away from trouble rather than towards specified goals, and (4) choosing among a few familiar alternatives only.

One should not necessarily imagine every single incremental step in the planning process proposed by Lindblom as separately decided by an Arrow type aggregation device. Lindblom is more preoccupied with the merits of a sequence of small steps, the one correcting the former and adjusting the outcome of the entire disjointed process. He is saying that "where analysis and policymaking are serial, remedial, and fragmented, political processes can achieve consideration of a wider variety of values than can possibly be grasped and attended to by any one analyst or policy-maker" (Braybrooke and Lindblom, 1963, p. 138). This favourable view of the whole sequence above the single decisions follows from Lindblom's limitation of the analysis of each incremental step so that "important affected values are neglected" (Lindblom, 1959, p. 81). Not only are unimportant effects neglected; the analysts often "rule out of bounds the uninteresting (to them), the remote, the imponderable, the intangible, and the poorly understood, no matter how important" (1959, p. 90). As restrictions are imposed anyhow, one might as well dismiss effects according to a criterion that offers some advantage in the subsequent decision process. One such criterion is to leave out impacts making it difficult to order the alternatives along a single issue dimension. Consequences are related to goals, so a restricted set of consequences can have the same effect as a limited goal-set in making evaluation rest on a single dimension, thus creating a basis for relaxing the condition (U).

Incrementalism relaxes (U) by restricting the reference group assessing each small planning decision. Restricting the set of values and consequences and limiting the set of decision-makers are two sides of the same coin. There are good reasons for Lindblom's refusal to use any more inclusive group as a steady reference. He realizes that all the difficulties of evaluation—the multiplicity, transitoriness, and conflict of values—are aggravated by shifting to larger reference groups (Braybrooke and Lindblom, 1963, p. 244). The recommendation of a sequence of small partisan decisions greatly facilitates the reduction to a single issue dimension for each separate decision.

Disjointed incrementalism shows a greater analytical preoccupation with ills to be remedied than with common goals to be sought. Critics like Etzioni (1967, p. 387) find that incremental steps "may be circular—leading back to where they started, or dispersed—leading in many directions at once but leading nowhere". It is far from incrementalism to opt for change at any price; this would be contrary to the spirit of the entire strategy. Nevertheless, when no goals have emerged and the feasible directions of change seem equivalent, a sensible issue dimension is status quo/change. Alternatively it can be termed status quo/risk, since all change involves risk, and substantial change is more risky. With only one dimension, cycles are prevented when choosing a way out of present hardship. Even when cycling does occur, its costs need not outweigh its advantages, as argued by Pildes and Anderson (1990, p. 2185): "(C)ycling might permit acknowledging inconsistent preferences as a way of expressing the incommensurability of certain goods. And cycling may well promote both stability and fairness, insofar as it enables rival majorities each to have their way for a time."

Riker (1982, pp. 244–245) sees the essence of voting in a democracy as the possibility for an electorate to remove officials from office. In this respect the collective choice procedure can be effective even if voting outcomes are ambiguous or unstable. All the alternatives in the new cycle may be better than the stable initial situation. One can defend fair procedure giving unstable results in a similar way under disjointed incrementalism. Even if the preferences are such that an aggregation device cannot prevent cycles, the alternatives comprised by the cycle may all alleviate present hardship. Moreover, when the lack of specified objectives reflects an absence of strong preferences, unstable outcomes may not be felt so hard.

Besides, an initially unstable outcome may be stabilized when posterior rationality is at work—and in disjointed incrementalism it is supposed to be.[1] Not just any kind of ex post evaluation would help; the attractive psychological mechanism is what Festinger (1964) called cognitive dissonance. There is a tendency to upgrade ex post the preference for the option actually chosen in order to preserve the image of oneself as a rational person. A similar mechanism may be at work on the level of collective decision-making. Not being able to adopt a particular alternative—regardless of which one—may bring the decision-making body and the governance system into discredit. Even individuals or groups that voted against the chosen option may want to prevent this. There may be a common interest in making the collective appear rational, and not the victim of uncontrived decision cycles. It may also be that intransitivity is due to deficient knowledge about the alternatives. It is conceivable that sufficient knowledge may be gained by experience, so that ex post rationality helps to reduce the likelihood of intransitivity.

The next point to be discussed is the limitation of incremental analysis to a few somewhat familiar policy alternatives. Disjointed incrementalism shuns "any policies whose scope is such that if they miscarry, the evils will exceed the remedial powers of existing institutions...(I)t does require that insurance be in force for each step" (Braybrooke and Lindblom, 1963, p. 239). Such strong risk aversion will turn down many plans. Not only is the exploration of substantially divergent alternatives omitted, but also incremental alternatives that fall outside the familiar path of policy-making. It is clear from this that incrementalism prescribes a procedure restricting the number of alternatives and the range of options on the agenda compared with synoptic planning. Thereby a situation is created in which the condition (U) requiring unrestricted choice is relaxed. The predilection for the familiar reveals a metapreference for uncertainty avoidance:

> If there is enough potential instability, then the ultimate effect of tampering with the status quo is apt to be quite uncertain. If preferences are cyclical, most voters will have alternatives which they prefer to that status quo and which could receive the support of a majority in pairwise contest with the status quo; but if there are cyclical majorities, there will also be (many) alternatives which might be chosen which could leave the voters worse off than if he simply reconciles himself to the status quo. Thus, in situations of complexity with cyclical majorities across multiple dimensions, voters may be very reluctant to 'rock the boat', since there is no guarantee that they will be better off doing so. (Grofman and Uhlaner, 1985, p. 41).

It is interesting to juxtapose the incrementalists' insistence on the special position of the status quo with McKelvey's (1976) result that once transitivity breaks down, it completely breaks down (shown in a less demanding way by Kelly (1988, Ch. 4)). "(N)ot only is a Condorcet winner unlikely, but also, when one does not exist, *anything* can happen. There is no 'small' set of probable outcomes" (Riker, 1982, p. 187). On the basis of theoretical analysis in social choice theory, Riker concludes that all methods of amalgamation of individual preferences to a social choice involve illogicalities, which is the fundamental basis for disequilibrium. Majority rule is often in disequilibrium, and "wide swings in political choices are possible and expected. Topsy-turvy revolution is as certainly predicted as incremental change" (1982, p. 188).

This result does not seem readily compatible with Lindblom's assurance that most collective decision-making is in fact incremental. Disequilibrium of tastes and values does not prohibit incremental change, but it may well lead to more fundamental alterations. However, individuals do not act on the basis of private tastes and values alone. "They also respect and are constrained by institutions that are intended to induce regularity in society. And it is the triumph of constraints over individual values that generates the stability we observe" (Riker, 1982, p. 190). However, Green and Shapiro (1994, p. 116) hold the theory of structure-induced equilibrium to be yet "another illustration of a literature groping for realism without being rooted in empirical inquiry".

Even if decisive empirical evidence is still lacking, it seems to be a reasonable hypothesis that a number of institutional practices counteract cycles. They reduce the number of issue dimensions or alternatives and put constraints on agenda setters (Douglas, 1984; Shepsle, 1979; Niemi, 1983). I shall mention just a couple of examples from planning. Firstly, in some countries certain antiquarian interests are protected by law. In that way an absolute restriction is created that serves to reduce the number of issue dimensions. Any alternative which makes it to the set from which the final choice is made, will be good enough from the antiquarian point of view. This means, for example that 'preserving monuments of the past vs economic efficiency' will not easily arise as a main issue dimension.

Secondly, it is customary to work out plans—for example road plans—at different levels and decide on them separately. One plan may determine what highway links are needed in a district. Another one is focused on the alignment and standard for just one of those links. A more detailed plan determines the location of access roads to nearby houses, the design of junctions, etc., for a single link. This partitioning of the problem is affiliated with, and indeed frequently combined with, the technique of distinguishing between main alternatives and sets of more and more specified subalternatives. Both this procedure and the institutionalized hierarchy of plans simplify decision-making and render cycles less likely (Kadane, 1972; Feld and Grofman, 1988). At the survey level some questions and problems are postponed, which may limit the number of issue dimensions. Besides, at the lower levels some issues have been settled, which reduces the number of alternatives. In addition, the budgetary process often forces the division of road schemes into smaller projects decided on in different years. As the standard should be uniform over the entire road connection and the two end points are most often given, many options on economy and alignment design will be closed before each single project is treated.

We have seen that disjointed incrementalism relaxes (U) in various ways. This is to say that incrementalists lower Arrow's ambitions of fairness and democracy. The planning process may be considered fair enough, and far more democratic than in the synoptic case. But each choice set is limited to alternatives close to the status quo, so the decision itself is severely constrained. The incremental procedures imply a trade-off between fairness and logicality (transitivity). Because several of the relaxations of (U) arrange for single-peakedness, social decision cycles are counteracted correspondingly.

4.2. PROPERTIES OF DISJOINTED INCREMENTALISM MODIFYING THE INDEPENDENCE CONDITION

This section deals with features of disjointed incrementalism violating the pairwise-choice format. These are (1) the special part played by the status quo, (2) the interweaving of what is desired and what is feasible, and (3) the striving for agreement. In addition, I comment on Braybrooke and Lindblom's recommendation of 'census data'.

The desired familiarity of policy alternatives reflects the salient point of the strategy, viz. the virtues of incremental change (Samuelson and Zeckhauser, 1988). The special predilection for new situations close to the existing one violates condition (I) (Klemisch-Ahlert, 1991). Imagine that the economy/environment dimension is judged important, and that two promising alternatives, A and B, are situated on the scale as shown in Fig. 2.

In the pairwise comparison of A and B, when the current state is believed to be SQ1, B seems most attractive. However, during the planning process, hitherto unknown pollution from the storage of toxic waste from the factories in use, is disclosed. The status quo (SQ) is reinterpreted, accordingly, to be situated at SQ2. With the prejudice in favour of small change inherent in the strategy, A might now be chosen, even if the benefit–cost ratios of A and B are not affected by the reinterpretation. Hence, the choice between A and B is not independent of the status quo, which is a third, 'irrelevant' alternative. The violation of (I) is implied by the acknowledgement of the status quo/change dimension as significant.

The most important aspect of the small-change prescription of incrementalism is not that it restricts the number of alternatives, but that each single decision becomes less important. This is at the heart of disjointed incrementalism as a sequence of trials, errors, and revised trials. Such sequential decision-making is a strategy for social choice under conflict. The problem is broken up in smaller parts, so that partial decisions can be made once sufficient agreement is reached. Riker (1982, p. 128) asserts that even if it is not possible in general to make social decisions with the qualities desired by Arrow, one is more likely to succeed for relatively unimportant decisions. The problems are then less complex and it is easier to agree on a single issue dimension. Besides, there is less to be gained by manipulation, e.g. by generating a cycle. This can be done by expressing insincere preferences or by manipulating the agenda, but the outcome must be worth the time and effort of prospective losers. The satisficing procedure recommending the first acceptable solution discovered makes incrementalism vulnerable to manipulation by planners in charge of the agenda. They can increase the likelihood of seeing their favourite solution realized simply by making it the first one to be subject to impact assessment.

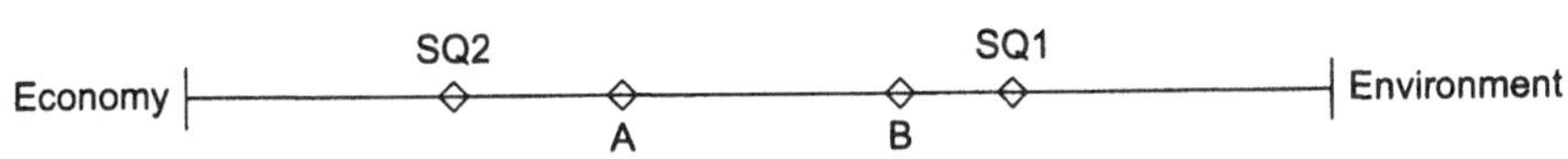

FIG. 2. Example of alternatives situated along an issue dimension.

However, satisficing also arranges for the use of procedures limiting the manipulation. I now assume that the option seeking continues until each chooser recognizes at least three acceptable alternatives. 'Approval voting' can then be practised (Brams and Fishburn, 1983). Roughly, everyone submits an unranked list of the alternatives of which they approve and the rule selects the alternatives appearing on most lists.[2] This seems as a reasonable procedure under incrementalism. Firstly, each chooser has an aspiration level yielding a set of feasible proposals. Secondly, the social choice rule ensures a high degree of agreement. We now call the preferences of an individual 'dichotomous' if there are exactly two indifference sets, e.g. one set of unranked alternatives that are acceptable and one other set of unranked alternatives that are not. The attractive feature of approval voting can then be stated (Kelly, 1988, p. 142):

> It has been proven that approval voting always induces sincere submissions when preferences are dichotomous. In fact, it is the only unranked voting system with this property. This is the sense in which approval voting meets our design specifications of minimizing strategic misrepresentation.

Nevertheless, approval voting sometimes fails to select a Condorcet winner even when everyone votes sincerely.

Manipulation is possible under any method of voting (Kelly, 1988, p. 172). According to Riker, no fair method of voting can satisfy the logical criterion of path independence. It would require that regardless of how alternatives are divided up for a serial decision process, the same choice will result. This requirement is especially relevant to incrementalism with its small-step sequential problem-solving. The final result can be influenced by the manipulation of the sequence of intermediate steps.

Path independence would be violated by a practice insisting that certain thresholds must be surpassed for change to make any difference (MacKay, 1980, Ch. 3). In transportation planning it has been argued for the use of thresholds when evaluating minor time savings (Bruzelius, 1979, pp. 183–189). The splitting up of road projects, each part yielding incremental time gains below the threshold, would affect the calculated profitability of the overall project and possibly cause decline in road construction. The acceptance of thresholds would thus make incrementalism vulnerable to path dependence.

'Staging' is an attractive form of dividing large projects into smaller ones that can provide benefits even before the later parts of the project are implemented. Path dependence is in fact wanted in staging. It is seen as a good thing that the assessment of the entire project depends on how it can be split into stages. Staging is a kind of incremental planning practice breaking the Arrow conditions because metapreferences are taken into account.

I now turn to the interdependence of means and ends in incremental decision-making. The condition (I) implies that the aggregation device ranks two alternatives without requiring knowledge of what other alternatives are in the choice set. Our desires are supposed to be independent of our opportunities; it is sensible to proceed with value calculations independently of feasibility considerations (Plott, 1976, p. 548). Disjointed incrementalism takes a radical departure from this basic assumption: ends are chosen that are appropriate to available or nearly available means. On the collective level preferences are adjusted to opportunities, and there is no independence of what Arrow called irrelevant alternatives. Incrementalists cannot collectively discuss what they would like to do before they are aware of what they are able to do.

Plott sees the denial of the value/feasibility distinction as the core of the criticism against Arrow's problem formulation: "Doing away with this is, in essence, doing away with the concept of a social preference definition altogether" (1976, p. 548). When preferences depend upon possibilities, when the ends depend on the available means, the preferences and thus the individual rankings lose their exogenous character. The individual orderings will then be susceptible to repeated change even if the set of alternatives is unaltered. The reason is that shifts in the view of the feasibility of the alternatives are permitted to affect the individual rankings. This paves the way for manipulation. However, the satisficing strategy of disjointed incrementalism points to a way out of the problem. The aspiration level can be set as standards, e.g. prohibitions against certain events. That does not presuppose the existence of a preference (or welfare) function, and it does not have to involve the concept of preference for social outcomes at all:

> That is, standards can look like *constraints*, e.g.: 'the outcome should not violate anyone's rights'; 'Thou shalt not kill'; 'No one should starve to death'; 'No Pareto dominated option should be chosen'; etc. So, an alternative to the use of a social preference definition is to use constraints. Rather than attempting to find what is *most* preferred or *most* acceptable, we can try to find something which is simply acceptable as opposed to something which is optimal. (Plott, 1976, p. 548)

Assume now that our aspiration level is identical with the limit of the feasible set. Then this set is empty until an alternative is found that we regard as acceptable. Satisficing will now save us from the Arrow paradox. We choose the *first* feasible alternative identified. There is no choice between feasible alternatives. The important decision is transferred to the choice of aspiration level. The satisficing principle can be modified to allow choice among the N acceptable solutions first found. Arrow's impossibility theorem becomes a troublesome reality again when $N \geq 3$. But even with three alternatives the probability of intransitivity is low

In the view of Plott (1976), the Arrow type paradoxes are caused by the fallacy of anthropomizing (i.e. the conception that the features of individuals will also characterize other phenomena, like groups, organizations, or God). In this case it means to illegitimately presuppose that a property of each individual in the group—viz. utility maximization—is also a property of the group. Maximizing a social preference function may not be a property germane to collective behaviour, and the same can be said about transitivity.

Criteria other than standards may be used to evaluate policy. Lindblom gives a prominent place to the following approach: "The test of a 'good' policy is typically that various analysts find themselves directly agreeing on a policy (without their agreeing that it is the most appropriate means to an agreed objective)" (1959, p. 81). Braybrooke and Lindblom (1963, pp. 137–139) hold that the same goes for policy-makers as for analysts. For each single step agreement is sought, not universally among policy-makers, but among those participating in the reference group in charge of the next small decision in the sequence. Planners could try to construct and make a survey of agreement, and then pick an action not expected to set off a fight. One acts according to anticipated reactions and in harmony with the pursuit of security and agreement in disjointed incrementalism. In predicting the degree of conflict, information on preference intensities is extremely useful. When planners employ such knowledge in their search for incremental change unlikely to meet resistance, the condition (I) is violated.

The predilection for consensus may be interpreted as a type of metapreference:

> (V)oters may believe that outcomes which have garnered some specified level of wide support should, as a matter of principle, be supported. Other voters may simply have low tolerance for disagreement. In either case, a voter's preference for consensus or for agreement may outweigh his or her preference for any specific outcome. (Grofman and Uhlaner, 1985, p. 41)

Lindblom thinks that such metapreferences tend to strengthen the position of the status quo. So do Grofman and Uhlaner (1985), as they argue that the metapreferences promote stability in social choice.

I now turn to the question of what kind of information should be accepted in incremental decision-making. A section is devoted to the question in Braybrooke and Lindblom (1963, pp. 177–193). Disjointed incrementalism requires commensurability of values and hence of consequences affecting the values (1963, p. 150).[3] Small adjustments of policy taking care of some values are moderated by other adjustments promoting another value-set. The commensurability makes mutual adaptations by partisan interests possible. Braybrooke and Lindblom (1963, p. 206) state that "several features of the strategy of disjointed incrementalism make it easy to associate the strategy with utilitarianism"—and commensurability is one such feature.[4] Utilitarianism is often accompanied by the belief that the utility of an action or policy can be reduced to the value of the caused pleasure and pain. This 'felicific calculus' (in the terminology of Braybrooke and Lindblom) would insert cardinal information into the evaluation of policies and hence violate condition (I). However, Braybrooke and Lindblom want to suspend this utilitarian calculus from their incrementalist approach. They prefer to drop data on the intensity, duration, etc. of pleasure and pain and replace it by information of the census type.

The census notion is concerned with the distribution of objective properties among persons, regardless of personal beliefs or preferences. It is a registration of how many in a certain reference group or population display the characteristics of, say, experiencing traffic noise above 60 dB(A) in their home, travelling more than 30 min by car to work, or perceiving the nearest main road as a barrier. Such census variables are easily combined with the setting of standards and thus with a satisficing criterion. One should predict such census type variables for the small projects under consideration, and choose an alternative with a sufficiently promising after-situation. Interpersonal utility-comparisons would then be eliminated, as one person's condition is counted for no more and no less than any other's:

> A census requires interpersonal comparisons (of a kind constantly made in everyday life), but it does not require a comparison of utilities or an interpersonal calculus of any sort, that is to say, any method of assigning numerical scores or measurements to the persons surveyed, or to their properties. It neither presupposes such a calculus nor does it exclude the possibility of one. (Braybrooke and Lindblom, 1963, p. 181)

Interviews are in many cases required to obtain census data. Properly designed stated preference technique can strongly reduce the likelihood of strategical answers (Pearmain and Kroes, 1990). Census data are liable to conceal some other problems, however, and Braybrooke and Lindblom mention the "mixed changes" and the "change of persons". The first case occurs (for example) when a policy enlarges both the stratum experiencing good conditions and the stratum enduring bad conditions of, say, health (when there is also a middle condition). The second case occurs when, although the percentages of the strata are such as to indicate an improvement, a lot of those in the well-off group ex ante are to be found in the badly-off group ex post. Then there is a case of victimizing some people in order to benefit others.

These effects suggest that there can be a consensus in favour of one policy even though the census data indicate that a large majority supports another policy. The use of census data can thus lead to violation of the Pareto condition (P). Imagine three persons, A, B, and C, and the three variables noise, safety, and driving expenses—each with a high, middle, and low category for all three variables. In the example, A and B will be exposed to less noise, B and C will experience a safer neighbourhood, and A and C will have their driving expenses cut. However, A suffers reduced safety and thinks this outweighs the two advantages. B will incur increased outlays, which means more to him than the incremental benefits in safety and environmental standard. C will be exposed to more noise, which he deplores more intensely than he appreciates the monetary and safety gains. Now, for all the three census variables, two thirds of the reference group are registered as having attained a better situation. Nevertheless, A, B, and C did not approve of the proposed policy. In a community where solidarity is valued, the same result could occur because the gainers might oppose alternatives causing significant loss to some people, even if they are few.

Braybrooke and Lindblom (1963, p. 228) hold disjointed incrementalism utilizing the census notion to be an effective, though not invariably successful, replacement of Bentham's utility calculus. As compared to that 'felicific calculus' and to Arrow's ordinal, rival approach, theirs is in fact possible in practice, and "may shine by contrast", as they say (1963, p. 193).

4.3. COMPARING SYNOPTIC AND INCREMENTAL PLANNING AS STRATEGIES FOR ALLEVIATING THE EFFECTS OF THE SOCIAL CHOICE PARADOX

How different are really the synoptic and incremental modes of planning? Planning theorists have diverging viewpoints on this because they want to achieve different things. Sager (1995) designs the two modes as antipodes because the aim is to highlight some 'ideal types' for mapping the terrain of planning theory. Weiss and Woodhouse (1992) soften the extremities of disjointed incrementalism in order to encourage both critics and defenders to get on with the task of learning more about how individuals, organizations, and societies can proceed relatively intelligently despite the fact that humans rarely have a good understanding of complex social problems and policy options.

One disputed issue concerns the incrementalists' alleged conservatism and the corresponding favouring of organized elites. Admittedly, there is something of a rigged competition in which business and other well-organized interests win disproportionately. Weiss and Woodhouse (1992, p. 262) hold, however, that there "would be obstacles to marshalling political resources behind *any* change-oriented strategy. Hence, it is the socio-political power structure that favours organized elites, not incrementalism or any other decision strategy".

One might argue, as Lindblom (1979, p. 520) does, that "(a) fast-moving sequence of small changes can more speedily accomplish a drastic alteration of the status quo than can an only infrequent major policy change". Lindblom contends that incrementalism is usually the available method leading to the fastest change, because the small steps "do not rock the boat, do not stir up the great antagonisms and paralyzing schisms as do proposals for more drastic change" (1979, p. 520). In other words, throughout the political process, the status quo can refer to a succession of rapidly changing states. This line of argument weakens the special position held by the status quo in incremental planning. Significant alteration may be accepted over a short period of time as long as errors can be effectively corrected. Besides, no strategy for public planning can completely disregard *in practice* that the likelihood of realizing a specific proposal is affected by its 'distance' from the pre-existing situation. Neither can the need for substantial political support be ignored in practice.

One cannot really evade the Arrow paradox by simply relaxing one or more of his assumptions. The paradox continues to be troublesome unless nothing important is given up by the relaxations. The suggested qualifications imply a slackening of the demands made either on fairness or on logicality. I have not proved that the violations of (U) and (I) inherent in the synoptic and the disjointed incremental planning are really effective in constructing a social preference ranking that observes the rest of Arrow's conditions. What I have offered is only the barest sketch of how to alleviate the effects of the paradox in ways suggested by the planning procedures.

A summary conclusion of the discussion of synoptic and incremental planning could go as follows: Synoptic planning arranges for rational social choice by violating both the ordinality aspect and the pairwise-choice format of the condition of independence of irrelevant alternatives. Disjointed incrementalism attacks condition (U) to a far greater extent, exhibiting a number of attributes restricting the scope of analysis and decision. This is not surprising as a main conclusion, since (1) it is a central feature of synoptic planning that it requires extensive information, and (2) it is essential to disjointed incrementalism that it simplifies the decision situation. Nevertheless, the rejection of the value/feasibility distinction and the stated preference for agreement and solutions adjacent to the status quo, also make incrementalism violate the pairwise-choice format of the independence condition (I).

Public planning should assist the government and elected politicians in making informed social choices. Both the synoptic and incremental approach aim at providing such help in a rational way, however differently they define 'rational'. One requirement concerning rationality, which can be set for both strategies, is that the recommended decision-making process should not be too discredited by Arrow's impossibility theorem. Generally, it is impossible to ensure that the results of collective decisions will be meaningfully related to the preferences of citizens. I have demonstrated that synoptic and incremental planning have properties tending to create situations in which it is not unreasonable to relax one or two of Arrow's conditions, so that the avoidance of cycles in social choice is theoretically possible. Most often the condition of unrestricted scope (U) or the independence of irrelevant alternatives (I) is relaxed, retaining the attractive democratic qualities of Pareto-optimality and non-dictatorship. The ways in which the two planning modes relax one or more of the conditions for a satisfactory aggregation of individual preferences enable us to better understand how planning in general functions in the political system. Public planning propounds institutionalized ways to arrange for social contexts in which the logicality of political decision-making processes can be achieved, while maintaining at least a minimal degree of fairness. It does so by

—providing a knowledge-generating process making it less dubious to employ certain types of other information in addition to preference rankings,
—discouraging certain types of manipulation,
—increasing the likelihood of single-peaked preferences,
—and by simplifying the decision situation in other ways.

Nearly perfect information leaves little to be gained from manipulation and persuasion. The combination of expert-oriented synoptic planning on this basis and an Arrow type of social preference ranking founded only on the individual orderings leaves practically no room for politics. On the other hand, disjointed incrementalism is the planning process of a pluralistic political system. Miller (1983) has pointed out that the preference patterns identified by pluralist theory as promoting stability, are essentially those identified by social choice theory as entailing instability.[5] The indeterminate direction of the sequence of future incremental steps with its inherent possibility for cyclicality may counterintuitively contribute to the stability of the pluralistic political system: "Precisely because social choice is *not* stable, i.e., not uniquely determined by the distribution of preferences, there is some range for autonomous politics to hold sway, and pluralist politics offers almost everybody hope of victory" (Miller, 1983, p. 743). It seems reasonable to me that widespread acceptance of the existing constitutional arrangements—the loyalty to pluralistic democracy—is ranked above the desire for transitivity of social preferences. Then it is a question of how far one should go in the direction from incrementalism to synoptic planning in order to do away with some of the indeterminateness and thereby achieve transitivity, e.g. by bowing to the cost–benefit analysis.

Planning should help to prepare for a good decision. Part of the preparation might be to alarm the public when decisions are not as good as required, for example when the outcome is merely an artefact of the decision procedure rather than an expression of the public opinion. Radcliff (1992, p. 47) asserts that there would be no difficulty, were it known when the General Will had so degenerated. The arbitrary decision would simply be declared null and void. He suggests a research programme aimed at effective and easily institutionalized methods for exposing arbitrary outcomes when they occur (1992, p. 48).

NOTES

1. Posterior reasoning is intrinsic to the incrementalists' way of thinking. March (1978, p. 593) gives a brief outline of the thought process: "Ideas of posterior

rationality emphasize the discovery of intentions as an interpretation of action rather than as a prior position...Actions are seen as being exogenous and as producing experiences that are organized into an evaluation after the fact. The valuation is in terms of preferences generated by the action and its consequences, and choices are justified by virtue of their posterior consistency with goals that have themselves been developed through a critical interpretation of the choice. Posterior rationality models maintain the idea that action should be consistent with preferences, but they conceive action as being antecedent to goals."

2. Kelly (1988, p. 142) defines approval voting as follows, where X denotes the choice set: "The unranked voting system of *approval voting* is defined as follows: First, all non-empty subsets of X are permitted; one such subset, S_i is submitted by each individual, i. Then, the system chooses the set S of all those alternatives x such that, for all y, the number of S_i in which x appears is at least as large as the number in which y appears."
3. Braybrooke and Lindblom distinguish peremptory and meliorative values, and the last mentioned are what I mean by commensurable values: "On a meliorative approach, judgments about accepting or rejecting any policy must wait upon a comparison of that policy with alternatives to it. On a peremptory approach, certain characteristics are looked for on the basis of which a policy would be approved or disapproved taken by itself, without any attention necessarily being given to alternatives. Clearly, the margin-dependent feature of the strategy makes it meliorative rather than peremptory, for it requires that the evaluation of any given policy reflect the marginal differences between that policy and alternative ones; and these differences cannot be discovered without comparisons." (Braybrooke and Lindblom, 1963, p. 150)
4. Braybrooke and Lindblom delimit utilitarianism in the following way: "*(A)ny* theory belongs to the utilitarian family if, first, it allows meliorative and distributive considerations a decisive role in confirming or disconfirming all moral judgments that are subject to dispute; and if, second, it supposes that, among these meliorative and distributive considerations, social welfare and group happiness justify supporting an action or policy, while their opposites do not." (Braybrooke and Lindblom, 1963, p. 206)
5. Pluralist theory sees society as divided along one or more lines of fundamental conflict or cleavage that partition its members into different groups. The preferences of members of society, with respect to alternative public policies, are largely determined by the group to which those members belong. Controversies in pluralistic societies are likely to be judged along multiple issue dimensions. Pluralists contend that the available evidence suggests that the chances for stable democracy are enhanced to the extent that groups and individuals have a number of cross-cutting, politically relevant affiliations (Miller, 1983, pp. 735–736).

CHAPTER 5

Social choice and communicative planning

The planning theories compared in the preceding section are founded mainly on instrumental rationality. There is at least one other type of rationality that is important in planning, namely communicative rationality as defined by Jürgen Habermas. After some remarks on rational planning and power in light of social choice theory in the first section, the meaning of communicative rationality is briefly outlined in the second section. It is explained there how communicative rationality and the striving for dialogue in planning can make it less serious to modify some of the Arrow conditions. The last section discusses how conflict and conflict management techniques imply such modifications. In the entire chapter, the aim is to discuss features of communicative planning justifying a weakening of some Arrow conditions, thereby opening for avoidance of the social choice paradox. Communicative rationality is a central concept in the recently developed mode of planning called critical pragmatism (Forester, 1989, 1993; Sager, 1994).

5.1. INTRODUCTION ON RATIONALITY AND POWER

Planning theory is concerned with how to prepare for rational social choice. However, as social choices will often be disordered or inconsistent as long as society preserves its democratic institutions, it might be asked if rational planning is a meaningful term? Faludi (1987, p. 4) conceives of rational planning as a rule for testing decisions. It is a methodological rather than a behavioural rule. When subscribing to this view, planning can be rational even if social decisions are not. What might grow out of the paradoxes of social choice theory is not a demonstration of the futility of rational planning, but rather an altered and extended role for planners. When there is a cyclical majority for a number of the alternatives investigated, would it be illegitimate for planners to act politically? That is, should they openly act on values and beliefs as well as strict expert knowledge to try to break the cycle at the alternative receiving the highest score in the professional assessment? Why not work out a broader basis for expert intervention by wiping out the traditional distinction between politics and research when majority decisions are found to be illogical or arbitrary? What good will the detached, neutral, scientific planner do when participants have most likely been manipulated, when the public interest is not defined, and the political outcome is unstable and liable to alteration based on the power of group interests?

The outcome of a discourse on these matters may be less respect paid to the elected politicians directly experiencing the curse of the Arrow paradox, and more prestige to the expert trying to cut through the 'democratic confusion'. In short, the paradoxes of social choice may grant cheap points to technocracy, pushing planners further into the political sphere.[1]

Arrow's paradox reduces the possibility of backing up the calculated and technocratic solutions by referring to the public interest. The social choice paradox threatens to dry up the sources of one of the planners' traditional and most effective legitimations. A seemingly simple definition of 'the public interest' is 'what leads to a higher level on the social preference ordering'. The concept of public interest is then—in the welfare economic model—dependent on the possibility for aggregating individual preferences. Trouble arises when we demand that individual preference orderings are to be taken into account no matter how they look. The public interest would be a more operational notion in a society where the political institutions produced, say, only one issue dimension in each decision-making process. This would be a more likely outcome, for example, when the democratic institutions limit the number of issues and the choice set. However, Arrow's paradox of social choice "leads us to suspect that no such thing as the 'public interest' exists, aside from the subjective (and hence dubious) claims of selfproclaimed saviors" (Riker, 1982, p. 137).

It might be different when debate, civic education, and political socialization have created some social homogeneity giving a high likelihood of single-peaked preferences. In such a 'value community' it does not have to be serious that the aggregation device restricts the range of permissible individual preference rankings.

Imagine a pluralistic and democratic society in which political thinking is not concentrated on a few values and power is not concentrated on a few hands. Neither community nor concentrated power are present to guarantee consistency in social choice.[2] In spite of this one does not renounce any of Arrow's fairness conditions and has to pay in the form of irrationality in collective decision-making. This opens the way for manipulation as pointed out by Riker (1982, p. 137; Riker, 1986). Social choice can be influenced by agenda control, strategic voting, and manipulation of dimensions (as exemplified in Chapter 6). Riker believes that strategic change of issue dimensions is the most frequently attempted manipulation device. The reason may be that "(t)his manipulation works even though those who are manipulated know they are being manipulated because, once a salient dimension is revealed, its salience exists regardless of one's attitude toward it" (Riker, 1986, p. 151).

No matter which strategy is chosen to arrange for collective decisions, it seems that instrumental power relations will be at the core of social choice.[3] One can either ignore some expressed preferences, or

—indoctrinate and force into line (breaking U);
—accept dictatorial decision-makers (breaking D);
—suppress the common will of people (breaking P); or
—open for manipulation (breaking I or U).

However, the introduction of 'value community' above indicates that social choice can be meaningful and fair even in some non-coercive circumstances. I will now specify communicative rationality as a contrast to means–end (instrumental) rationality and introduce communicative power as the opposite of instrumental power.

5.2. DIALOGUE AND SOCIAL CHOICE

The principal dividing line between types of rationality in planning distinguishes the type appropriate for goal-oriented behaviour within a means–end structured problem area, and models of consistent reasoning when means and ends are not distinct categories. The first kind has been recognized for a long time (Brubaker, 1984), and is discussed by Habermas (1971) as 'instrumental reason'. It tells us how to combine the means and use them to achieve given ends.

The identification and analysis of noninstrumental types of rationality have mainly gathered headway after the Second World War. Paul Diesing (1950, 1962) and Jürgen Habermas (1984, 1987) are among those giving significant contributions. Habermas (1987, p. 137) distinguishes three aspects of communicative action: "Under the functional aspect of *mutual understanding*, communicative action serves to transmit and renew cultural knowledge; under the aspect of *coordinating action*, it serves social integration and the establishment of solidarity; finally, under the aspect of *socialization*, communicative action serves the formation of personal identities."

Communicative rationality guides communicative action. It is found in speech meeting the validity claims of comprehensibility, truth, rightness, and sincerity, and at the same time aiming at mutual understanding and agreement. Within a speech situation thus defined, a community can rationally derive the goals to be collectively pursued. Values and norms, which could not be seen to have any rational founding under instrumental reason, may come into existence in a communicatively rational manner.

The coordination and socialization aspects of communicative action come close to what Diesing (1962) denotes "social rationality". Put in Habermas's phrases, the integrative aspect serves to avoid that the resources "social solidarity" and "ego strength' become scarce. It guides the part of communicative action aiming at integration on the personal and the small-group level. That is, it leads to integration which furthers personal growth and community.

In one way or another, a purpose or a goal can be formulated for any action. The means–end scheme is invalid in communicative rationality not so much because goals are too unclear and ambiguous to inform action—which they often are. It is more to the crux of the matter that the ulterior end is embedded in the activity itself. Dialogue, close ego-confirming relationships, and the experience of being able to make a difference when issues are discussed (democracy) are important to the development of mature personalities. Hence, they have intrinsic value independent of any goal-oriented strategy. Dialogue in Habermas's sense is motivated by no compulsion but the force of the argument itself. An argument concerning values or norms is valid or invalid without reference to operational ends. The introduction of objectives would simultaneously induce an irrelevant compulsion on the discourse and transform the dialogue to technique. It is, however, the decisive criterion of a dialogue—

as of integration—that it is *not* technique; and this is what invalidates the means–end scheme. Dialogue is communicatively rational conversation.

The concept of 'relational goods' was recently introduced in the literature on rational choice (Uhlaner, 1989). Relational goods cannot be acquired by an isolated individual; they are the fruits of relationships with others. Examples are social approval, solidarity, friendship, the desire to maintain an identity, and the desire to experience one's history (Uhlaner, 1989, p. 255). The pursuit of relational goods is guided for a large part by the integrative aspect of communicative rationality, although one may have instrumental motives too. It is unsatisfactory to deal with relational goods within the traditional means–end scheme, especially when ends are changed in the process of consuming such goods.

Hannah Arendt (1970, p. 44) provides us with a communicative concept of power closely linked to communicative rationality. "Power corresponds to the human ability not just to act but to act in concert. Power is never the property of an individual; it belongs to a group and remains in existence only so long as the group keeps together". Arendt understands power as the ability to agree upon a common course of action in unconstrained communication. Power is a collective effect of speech in which reaching agreement is an end in itself for all those involved. The communicative concept of power is, thus, not instrumental but presupposes communicative rationality. "(P)ower, far from being the means to an end, is actually the very condition enabling a group of people to think and act in terms of the means–end category" (Arendt, 1970, p. 51). It serves to maintain the praxis from which it springs. That is, the advantages of being able to go unanimously in for a common course of action serve to strengthen undominated communication as the praxis for reaching collective decisions.

It is relevant to assess Habermas's ideal of undistorted communication as a social choice model because individual preferences are to count. One immediately sees that it satisfies conditions (D) and (P). It does not, however, comply with (U) as one cannot make a collective decision on the basis of whatever set of individual orderings. Individual preferences which imply the domination of other people are not tolerated in undistorted communication. The decision reached must be according to impartial criteria and not reflect group interests. If special groups generally benefit, one will judge the dialogical model to be unfit in the prevailing political situation, as communication cannot be sufficiently freed from domination. Habermas's model also violates the (I)-condition, just like conversation not subject to special rules. The common search for mutual understanding does not preclude interpersonal comparisons. On the contrary, such comparison seems to be implied by the demand for empathy. Surely, only the power of the best argument is to count. But pointing out that a project will lead to big differences in economic welfare may be a good argument in itself.

Communicative rationality facilitates fair and transitive collective choice by relaxing (U). When integration is successful, a community is established which is sufficiently unitary to yield a single issue dimension when collective decisions are to be made. The ambition in the critical pragmatism approach to planning is even higher, however, as one aims to eliminate socially unnecessary distortions (Forester, 1989,

Ch. 3; Sager, 1994, Ch. 8). A planning theory uncompromisingly based on communicative rationality is outlined in Sager (1995).

The above points were set forth in Sager (1990, pp. 417–418), but the possibilities for attenuating the trouble caused by the social choice paradox by opting for a dialogical decision process is analyzed in more detail by Miller (1992). Some of his ideas are delineated in the following. I start with a central argument:

> The process of reaching a decision will also be a process whereby initial preferences are transformed to take account of the views of others. That is, the need to reach an agreement forces each participant to put forward proposals under the rubric of general principles or policy considerations that others could accept. Thus even if initially my aim is to support the claims of a particular group to which I belong or which I represent, I cannot in a general discussion simply say 'I claim that group A—farmers, say, or policemen—should get more money'. I have to give reasons for the claim. These might be that the group in question has special needs, or that it is in the common interest to improve the living standards of the group. By giving these reasons, however, I am committing myself to a general principle, which by implication applies to any other similarly placed group. Thus I am forced to take a wider view, and either defend the claim I am making when applied not only to my group but to groups B, C and D, which are like A in the relevant aspects, or else to back down and moderate the claim to something I am prepared to accept in these other cases too. (Miller, 1992, p. 55)

It should be noted that not only narrowly self-regarding arguments are likely to be withheld or corrected in public dialogue. The same will happen to preferences that are repugnant to the moral beliefs of society, for example racist views. Of course, also preference orders that are irrational because they are based on false empirical beliefs are likely to be modified.

Communicative rationality relies upon a person's capacity to be swayed by rational arguments and to lay aside particular interests and opinions in deference to overall fairness and the common interest of the collective. When people are to some degree communally orientated in their outlook, the mechanism in the quotation makes it less pressing to insist, as Arrow does, that *any* coherent set of individual preference rankings be handled by the aggregation device. Miller suggests instead "the weaker requirement that the social decision procedure should be able to cope with all possible sets of *post-deliberation* rankings" (1972, p. 60).

Miller claims that it is a virtue of deliberative democracy that it will reveal when the issue under discussion amalgamates separate dimensions of choice to which different voters attach different weights. When such a multi-dimension problem shows up, he argues that it may be possible to split the original decision into components if the original alternatives are not discrete and irreducible. The idea is to achieve one-dimensional components giving single-peaked orderings. Putting the bits together again, we would have an overall result which can fairly be said to represent the will of the majority (Miller, 1972, p. 65). The point here is not to change the preference orders, but rather to change the decision agenda so that non-cyclical majorities emerge.

The conclusion of this section is that communicative power in Arendt's sense can make it reasonable to relax Arrow's condition of unrestricted scope (U) and thereby create possibilities for meaningful and fair amalgamation of individual preferences. It is a main aim of planners working in the critical pragmatism mode to build communicative power.

5.3. THE ARROW CONDITIONS AND CONFLICT MANAGEMENT

The section starts with a brief outline of a couple of conflict types that have clear effects on social choice. Thereafter, some forms of communicative conflict management are discussed.

'Affective conflict' describes the situation in which at least one of the contenders holds as an independent choice criterion that the opponent is not to have his way. The predilection for alternatives in due distance along the issue scale from the alternative ranked highest by the adversary, violates condition (I). Hence, there is a type of conflict inherently incompatible with the Arrow conditions. The reasoning is parallel to that in the previous chapter, arguing that the incrementalist sympathy for small changes of the status quo violates the (I)-condition. Imagine three alternatives, of which x is known to be top-ranked by group B. Group A compares the other two alternatives, y and z. If either y or z is situated close to x on the issue scale, A—being engaged in an affective conflict with B—might choose the one that is not. A's ranking of y and z depends on the preferences of their opponent for the x-alternative.

Zero-sum conflicts will have an effect similar to the affective ones, in that an alternative which is more attractive to B would be less attractive to A. But no new choice criterion is added to the ordinary utility maximization. Information about B's preferences or about x is without interest to A in choosing between y and z, so condition (I) is not violated. Zero-sum conflicts usually imply the acceptance of a common issue scale. It would be sheer coincidence if, for each pair of alternatives, A should gain along the investment/safety scale exactly what B loses along, say, the time savings/aesthetics scale. Therefore, under zero-sum conflict there will be single-peaked preferences, and amalgamation of them to a social choice is possible.

Next, I examine whether the negotiation strategies are of any relevance to the social choice problem. Arbitration is contrasted to facilitation and Cormick's (1982) mediation approach. The arbitrator is asked by the parties to negotiate in order to settle the conflict, and the mandate may indirectly allow him to violate any of the Arrow conditions. If the ambitions of the arbitrator are that the parties respect the settlement and that only lenient forms of authority and persuasion prevail, he should probably respect (P) and (D) in order not to raise aggression.

There are several forms of assisted negotiation. Susskind and Cruikshank (1987) deal with facilitation and nonbinding arbitration besides mediation. Facilitation is a conflict management technique that can be given new contents in a planning process aiming at dialogue. The facilitator focuses almost entirely on process and "uses whatever tools are available to create and foster an environment conducive to joint problem solving" (1987, p. 152). The task is not only to take care of the practicalities, like arranging meetings and ensuring that notes and minutes of the meetings are kept. The facilitator establishes contact between the parties, gets them to the table, and sometimes even acts as a moderator.

Susskind and Cruikshank state that the facilitator's emphasis is on monitoring the quality of communication. The facilitator intervenes with questions designed to enhance mutual understanding. Discussion approaching Habermasian dialogue is very demanding, and it is probably necessary with an impartial intermediary checking that commonly agreed rules are observed. Are all parties given the same opportunity to set forth their points of view? Are the real goals and intentions openly stated? Are arguments backed by reason or by power? The facilitator has the difficult task of looking after the ideal role taking: he should find ways to make the disputing parties show empathy towards the problems and worries of their contenders. This is at the core of dialogical planning. The main point in a social choice context is that facilitating dialogue increases the likelihood of policy preferences that are single-peaked, as argued in the former section.

Mediation dissociates itself from coercion and brings communicatively rational elements into the negotiation process. An unequivocal standpoint is taken regarding the use of instrumental power. The mediation process recommended by Cormick and discussed by Amy (1987) is reasonably clear as to how an arrangement should be negotiated, what should be meant by an agreement, and what kinds of influence technique can be tolerated. Cormick (1982, p. 3) describes mediation as follows:

> Mediation is a voluntary process in which these involved in a dispute jointly explore and reconcile their differences. The mediator has no authority to impose a settlement. His or her strength lies in the ability to assist the parties in resolving their own differences. The mediated dispute is settled when the parties themselves reach what they consider to be a workable solution.

According to process-oriented planning theory, the conflict management process is as essential as the terms of the agreement. This is in accordance with the mediation approach of Cormick, underlining that "(t)he basic ethical principle that the intervener should espouse is self-determination for all parties to a conflict" (Cormick, 1982, p. 1). This requires a pursuit of mutually derived and acceptable outcomes. Note that self-determination can result in the decision not to proceed with the attempt to negotiate a settlement or a solution.

Cormick's mediator cannot negotiate agreement irrespective of the range of conflicting preferences, and hence (U) does not hold. It is very likely that (I) is violated too, in order that the welfare of the challenging parties can be compared and bargaining made part of the mediation process. On the other hand, the mediator will not violate conditions (P) and (D), as that would be incompatible with the above concept of 'mediation'.

The management of destructive and manifest conflict in planning, requires communicative as well as means–end rationality combined into an integrative/instrumental method. One can conceive of the dialogical and integrative part of such method as aiming at a transformation of the individual preference structures so that amalgamation of preferences will be possible. The instrumental part combines the means so as to generate alternatives gaining wide support.

In an integrative process involving integrative/instrumental method preferences and goals are never taken at face value. The same goes for agreement and the declared motives for conflict. When the integrative part of the method is emphasized, it does not really matter if there exists a social choice function answering to the Arrow conditions, because his basic assumption is not accepted anyway: the individuals do *not* have well-ordered preference structures on issues requiring integrative method. The kind of logicality assumed by Arrow (transitive and consistent preferences) cannot be expected when it comes to these matters. And it does not seem urgent to stress the claim for transitivity of collective decisions made on the basis of the aggregation of elusive, superficial, and inconsistent individual preferences.

NOTES

1. What the planner might find tempting, is to recommend a decision rule corresponding to Arrow's conditions in cases where it does not lead to cycles involving top ranked alternatives. When troublesome cycles occur, an extra decision criterion is brought to bear. It might be the CBA criterion or Rawls's maximin principle, for instance. The trouble is that there are a number of possible criteria, and that cycles may arise when choosing among them. Much is at stake, since the criteria can give widely diverging results: the CBA criterion favours efficiency and ignores equity, while Rawls's maximin criterion puts all weight on equity and ignores efficiency.
2. Some kind of power concentration can hardly be prevented if a modicum of consistency in collective decisions is to be guaranteed: "(B)y using the logical requirement of acyclicity, rather than transitivity, it is possible to find social choice that satisfies all of Arrow's fairness conditions as well as the revised condition of logicality...(H)owever, something very much like dictatorship is required to guarantee...acyclicity...(A)cyclic choice requires a 'collegium' such that alternative a is chosen over b if and only if the whole collegium and some other persons prefer a to b. Thus, although a collegium cannot unilaterally impose a choice, unlike an oligarchy it can always at least veto." (Riker, 1982, p. 131)
3. Social choice theory can thus be taken to lend Friedrich Hayek some support: Hayek's (1944) entire idea of planning as the road to serfdom is based on arguments contending that planning implies concentration of power. Arrow adds that without concentration of power one cannot make the meaningful and fair social choices which planning aspires towards. In a way Arrow can be said to close the circle: many choosers and diverging preferences lead to claims for coordination and equality; this leads to central planning (Hayek), which in turn clears the way for the concentration of power and lack of freedom (Hayek), and this makes meaningful social choice possible even with many choosers and diverging preferences (Arrow). But you cannot at the same time have the logicality (transitivity) and the freedom—when freedom is understood as the fair consideration of everybody's individual preferences in collective choice (Arrow).

CHAPTER 6

How decision cycles arise: two cases

The purpose of this chapter is to illustrate the difficulties of aggregating preferences into something that can reasonably be called the will of the people, when no single party provides a clear majority in support of any alternative (Schwartz, 1981). The two cases deal with the siting of the new main airport in Eastern Norway and the gradual closure and succeeding reopening of tramlines in the City of Trondheim, Norway. The airport case is an exemplification of Condorcet's voting paradox and some variants of strategic voting. The tram case shows how the social choice can be affected by party structure and the manipulation of alternatives, issue dimensions, and preferences.[1]

The crucial difference between the cases is that the airport case illustrates a cycle with unchanged preference orderings, while the orderings changed throughout the political process in the tram case. Thus, the cycles over time in the tram case may be due to other mechanisms than the Arrow paradox. They nevertheless exemplify well some causes for cycling dealt with in the social choice literature.

6.1. THE PARLIAMENTARY VOTE ON THE SITING OF THE NEW MAIN AIRPORT IN EASTERN NORWAY

The siting of the new main airport occasioned a lengthy process of planning and politics. Since the Airport Committee was appointed in 1968, the matter has been more or less continuously debated for 25 years. Gardermoen and Hobøl were among the alternatives from the beginning of the process, but a considerable number of other locations have been studied since then. Hurum, south of Oslo, was in fact chosen by the Storting (the Norwegian parliament) in 1988, but rejected later on because it was found to be too vulnerable to fog. The new airport is currently being constructed at Gardermoen. Here, however, the subject is not cycles occurring over time but cycles arising within a single voting scheme during which the preference orderings are unchanged. The first part of the case in this section studies the final vote in the Storting on 8 October 1992.[2] The second part comments on strategic voting on the basis of the vote on airport siting in the Storting on 8 June 1988.

The case illustrates Condorcet's voting paradox of the cyclical majority. Even when present, the paradox is not often explicitly revealed in a voting situation in the sense that the complete rankings of all the parties are really known. In the present case the rankings seem fairly clear, although it cannot be strictly proved that they were as claimed here. The uncertainty reflects that the voting procedure does usually not require that the voters reveal their full rank-orders. The final choice is frequently for or against a particular option or between pairs of options in the choice set. Some alternatives are often disposed of in preceding committee work or by preliminary voting. Figure 3 shows where the sites being seriously considered are located around Oslo.

An important step in the convergence towards the final choice set was the discarding of Hurum in 1990. When the date arrived for the parliamentary debate of the air-

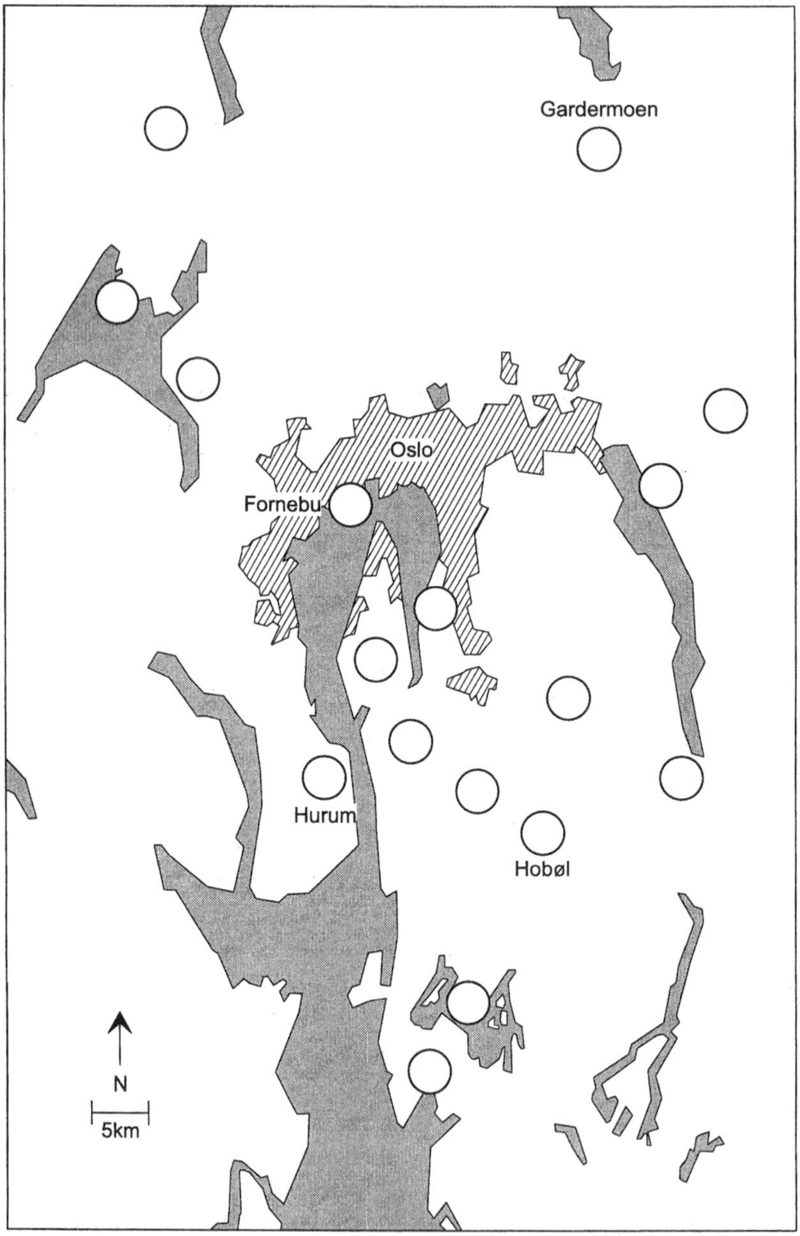

FIG. 3. Sites considered for the new main airport of Eastern Norway.

port siting, there were still five proposals, however. The Progress Party wanted a new airport financed, constructed, and operated by private business. The Socialist Left Party wanted to restrict air transport. Therefore, they proposed to close down the existing Fornebu airport and construct only one runway at Gardermoen, while the Labour Party proposal implies two runways. The Storting first voted over the proposition of the Progress Party, since the character of this alternative is basically different from the others. When this was defeated, they next voted over the proposition of the Socialist Left Party, since it was set forth by a smaller fraction of the Storting than the remaining proposals. Even this proposition failed to pass.

Following the elimination of the proposals mentioned above, the remaining choice set was:

G: Gardermoen
H: Closer analysis of Hobøl
S: Split solution with the domestic flights remaining at Fornebu (the current main airport), and the international flights using Gardermoen.

Three political blocks formed before the final vote. The Labour Party stood alone, while the Conservative Party and the Progress Party formed a coalition. The third block consisted of the Socialist Left Party, the Christian Democratic Party, and the Centre Party. The rankings of the three groups are shown in Table 1. The political blocks are denoted the g-group, the h-group, and the s-group after their top ranked alternatives.

If each representative follows the policy of his or her party when voting, the g-group controls 64 votes, the h-group controls 59 votes, and the s-group controls 42 votes. Thus, no single group is in a majority position but any pair of groups is. In this situation, the preference structure generates a voting cycle: Gardermoen will not be chosen because the h-group and the s-group prefer a closer analysis of Hobøl; further study of Hobøl will not be chosen, however, because the g-group and the s-group prefer the split solution; and finally, the split solution is beaten by Gardermoen in the rankings of both the s-group and the h-group.

In cases like this, the Storting votes according to a serial procedure. It can start by voting on the proposition considered most extreme or the one supposed to gain least support. The national assembly has to decide on the procedure, and in the airport case it was voted over two alternative successions. The President of the Storting proposed the voting order S, H, G, while the Parliamentary Leader of the Labour Party proposed the order G, S, H and was backed by a majority.

The Storting voted over Gardermoen (G versus not G). If G had not been chosen, there would have been a vote on S next. Since nearly everybody was of the opinion

TABLE 1. Group rankings of the proposed sites for the new main airport in the parliamentary vote of 8 October 1992

Ranking	g-group	h-group	s-group
1	G	H	S
2	S	G	H
3	H	S	G

that one of the proposals should be carried, the second vote would actually have been between S and H, and S would then have won. This result of a prospective second vote might be anticipated by the representatives, however. If so, they would regard the first vote as a choice between G and S. With a rational voting behaviour, then, a majority of the representatives would opt for G in the first vote. And this was, in fact, what happened.

To make my point, I nevertheless need to analyze what would have been the outcome, had the voting succession of the President been followed. The Storting would then have voted on the split solution first (S versus not S). If S was not chosen, there would have been a vote over H next. This second vote would be regarded as a choice between H and G, and Hobøl would then have won. Anticipating this result and acting rationally in the first vote, the representatives would see it as a choice between S and H, and a majority of them would vote for S.

It can be concluded that in the case of the new main airport in Eastern Norway, the voting order strongly affected the choice of site. Arrow's (I)-condition was violated, since the social choice did not depend only on the preference rankings of the choosers. The choice was not independent of the road leading up to it. However, this could not be prevented, as no site was backed by something deserving to be called the will of the people.

A similar theoretical voting analysis was published by Rasch (1990) on the Parliament's siting decision of 8 June 1988. That date the Storting made its Hurum decision, which was later to be reversed. The representatives were given a free hand to follow their conviction in the 1988 vote. Rasch does not know the preference rankings of a sufficient number of the representatives to conclude whether there was a Condorcet winner or a cyclical majority. Be that as it may; the 1988 vote is used here only to illustrate strategic voting.

The following citation by the representative Lars Velsand from the Centre Party reveals a clear case of voting for an alternative which is not ranked as the best in the choice set:

> In the final division I cast my vote for Hurum, with a minority of my group. I have to admit, though, that it is my clear conviction that *no* new main airport should be built. We should continue to use what we already have, i.e., Gardermoen and Fornebu...Following the defeat of the split solution in the Storting, the question arose what would most likely prevent a new main airport from being built, or at least delay it for a period long enough to warrant reconsideration of the entire project. The plans for Gardermoen had already reached such an advanced stage that I am convinced that the project would have started almost immediately. Hurum, on the other hand, was further out in the fog, literally speaking...The Hurum decision may turn out to give us a split solution after all, and thus a utilization of the resources and facilities already available to air transport. And the Hurum airport may not be built at all.

Besides exemplifying the main characteristic of strategic voting, the quotation demonstrates a particular feature, namely, how uncertainty may give opportunities for such voting. Moreover, Velsand draws on a problem often realized by the planners themselves; that is, the fragile link between planning and implementation. When Velsand regards the choice of Hurum as preferable to a Gardermoen solution at this stage, the reason is that his choice is made under uncertainty. The decision of the Storting concerns location, following which a planning process will be initiated. The real construction decision will be made on a later occasion. In Velsand's opinion, it would be easier to direct massive criticism against Hurum than Gardermoen, and he gambles that the objections will be strong enough to overthrow the project. Considering all that had happened in the main airport decision process so far, to him it did not seem too unrealistic to expect new alternatives, new issue dimensions, new political coalitions, etc. providing new opportunities for manipulating the decision process in the desired direction. The later development of the case, especially the rejection of Hurum and the decision cycle in the parliamentary vote of 1992, has shown that his strategy might well have been successful.

The 'free rider' variant of strategic voting to be exemplified in this paragraph does not depend on uncertainty about the future decision process. In the siting vote of 1988, there were several possible way to please one's own constituency while relying on others to bring about the airport site judged to be the best. For instance, some Conservatives might have been tempted to vote for Hurum to please the Royal Norwegian Air Force, even if they were of the opinion that Gardermoen would give the best airport. The Air Force would have serious difficulties with its operations at Gardermoen if that site was developed into a main airport. The Labour Party representatives from the districts around Hurum were probably under a heavier cross pressure, though. The party line was to support Gardermoen but the Hurum district also needed the activities generated by a main airport. Among the 14 Labour representatives voting for Hurum in 1988, there were—very likely—some voting strategically to please their local constituency and trusting others to carry through the party line. This time they miscalculated, however, and Gardermoen which was recommended by the Labour Government was outvoted.

6.2. THE TRAGIC TALE OF THE TRONDHEIM TRAM

The tramline system in Trondheim had a sound economy throughout the 1950s.[3] However, the rapidly growing number of private cars, the decreasing population density in the central parts of the city, and the scattered new residential areas further from the city centre were soon to change things for the worse. The term 'tragic' in the heading does not refer to the poor economic record in later years, however. Neither does it indicate a certain answer to the question whether Trondheim should have trams. From a social choice point of view, the tragedy of the decision-making process concerning the tram is that it erodes the constituency's trust in local politicians and thereby reduces the possibilities of finding public solutions to common problems in the future. Waste of public funds, virtually incomprehensible undecidedness, and arbitrary outcomes are all too clearly demonstrated in the tram case. This seems by far to overshadow the general argument of Miller (1983) and Pildes and Anderson (1990, p. 2185) suggesting occasional positive effects of cycles.

The purpose of including this story is to point at types of input to the political decision-process, which may cause cycles over a period of time. In this case, there are cycles in the political attitudes towards the tram: wanting to close down lines, deciding to keep the tramline system, actually closing it down, and allowing one line to be reopened.

The long-lasting discussion about the Trondheim tram took off in 1956 when the tramcar depot burned down and most of the tramcars were destroyed.[4] As the tram was good business at the time, it was quickly decided to buy new equipment and reopen the lines. There even was an extension from Voldsminde to Lade in 1958. Private cars were rationed in Norway until 1960. This helped to keep up the demand for public transport. Roughly speaking, there were four tramlines in Trondheim at that time: the private Gråkallbanen line (Lian–City Centre) and the municipal lines Singsaker–Railway Station, Ila–Lade, and Elgeseter–Voldsminde. The municipality bought the Gråkallbanen line in 1965, and later the new Lian–Lade line replaced the old lines to Ila and Lian. All the lines mentioned in this essay are shown in Fig. 4.

The optimism with respect to the tram did not last for a long time after the rationing of private cars ceased. In 1964 the City Council voted unanimously against extending the tramline further along the ever more heavily trafficked approach to the city centre from the south. Instead, it was decided, symptomatically, to widen the street with two new lanes. The gradual close down started in 1965, when the City Council decided to 'temporarily' stop the maintenance work on the Singsaker–Railway Station line. A few years of deteriorating service made it relatively easy to close down the line in 1968.

Throughout the 1970s there was a certain political consolidation of the position of the tramline system. There is a complex of reasons for this which I will not try to sort out. Two points should be mentioned, though. The first is that environmental considerations made their way into impact assessments of public projects and policies from the late 1960s onwards. The second is the increasing awareness of environmental qualities and living conditions in the cities, combined with an emerging critique of planning on the terms of the motorists. Anyway, the City Council voted in favour of keeping the two remaining tramlines both in 1975 and 1979. It should be added,

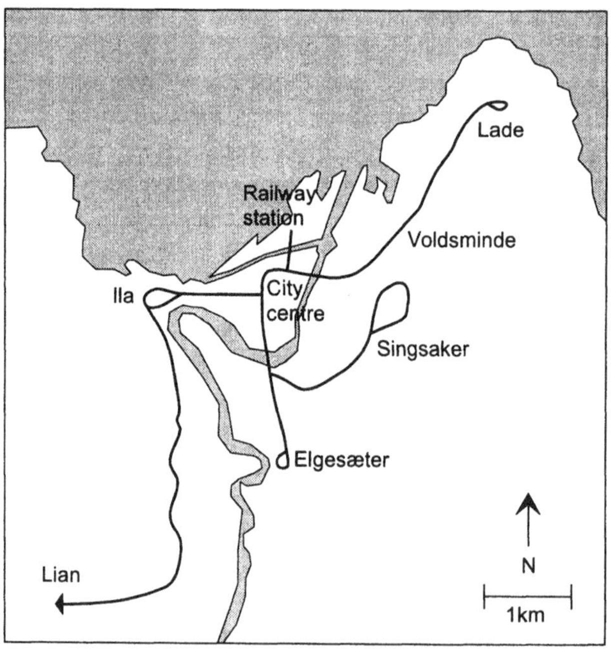

FIG. 4. Tramlines that have been operated in Trondheim.

though, that proposals for extensions were defeated on both occasions. As a consequence of the decisions in favour of continued operation, the City Council in 1982 voted for the purchase of 11 new articulated tramcars.

The political block supporting the tram and those opposing it were of about equal size in the City Council throughout the 1970s and 1980s. Therefore, when a few representatives in the pro-camp objected to the NOK 66 million investment in new trams (about £6.6 million), it was enough to create a majority for further contraction of the tramline system. In fact, the political constellation produced simultaneous decisions for investment in new tramcars and close down of the Elgeseter–Voldsminde line in 1982. These decisions in turn made it necessary to invest in a new NOK 28 million repair workshop and tramcar depot, as some of the old facilities were situated along the line to be shut down. Since the infrastructure had to be improved for another NOK 6 million, the total investments assumed to be required amounted to NOK 100 million (£10 million).

During the period of several years when the new tramcars were delivered and the new service facilities constructed, the pressure to get rid of the last tramline was upheld. Although the request for new studies of possible extensions and analysis of the future effects of total close-down succeeded in prolonging the process, in 1987 the City Council decided on the final close down of tram operations. The trams stopped running on the Lian–Lade line in June 1988. The last of the new tramcars were delivered early in 1985 but due to the cold and wet climate in Trondheim, adjustments had to be made over a fairly long period. The 11 new tramcars were stored away only about a year after being officially handed over from the producer.

Had the municipal public transport company been successful in selling the tramcars, the story of the Trondheim tram would probably have ended here. However, the sale of the 11 new tramcars turned out to be difficult. Trondheim's combination of narrow-gauge tracks, broad coachwork, and unidirectional tramcars seems to be unique in Europe. By 1990, a seriously interested customer had still not been found. Moreover, partly due to the recession in Norway since 1988, the new tramcar depot was not yet transferred to other use yielding revenue.

As it were, the closure did not signal the start of a peaceful period. Already later in 1988, the pressure group Gråkallbanens Venner (Friends of Gråkallbanen) passed over to the executive committee of the City Council their study on private reopening of the Lian–City Centre line. Furthermore, a new political party was organized in Trondheim before the local election campaign in 1987. The *raison d'être* of this City List (Bylista) is to fight for the tram. The City List's group of six in the City Council obtained their desired key position between the Labour and Socialist block on the one hand and the non-socialists led by the Conservative Party on the other. Hence, in the summer of 1989 both the political blocks advanced proposals accepting private reopening of Gråkallbanen on the condition that the municipality of Trondheim should not cover any kind of operational expenses. Following a discussion of the terms of hiring tramcars, depot, and infrastructure, private reopening was decided by the City Council in 1990, and the trams started running in August that year.

At the end of the two year trial period, the municipal administration concluded that the limited liability company Gråkallbanen had not run their business according to the agreed terms, and in the City Council a new attempt was made to close it down. The attempt failed, however, and the tram has since received increased subsidies from the Central Government. This seems to make the company capable of continued operation, at least until significant investments become imperative.

To make my decision theoretical points, I now start after the reconstruction following the fire in the tramcar depot in 1956. The positive period for the tram after the reconstruction ended with the rejection of extending the line southward from the city centre in 1964. I divide the period from 1964 onwards into four phases:

(a) Contraction and doubt (1964–1975)
(b) Confirmation of the *status quo* (1975–1982)
(c) Line by line close-down (1982–1988)
(d) Private reopening (1988–).

Our interest is now to see what input to the political decision-making process contributes to the transitions from one phase to the next. The three shifts are associated with decisions in the City Council expressing changes in the majority's political attitude towards the tram. In a social choice perspective, I comment on some of the mechanisms tipping the balance from a con-tram to a pro-tram majority or vice versa, thus facilitating the passage from phase to phase.

The closure of the Singsaker–Railway Station line led to a mobilization among the adherents of the tram. Throughout the 1970s prospective extensions of the tramlines were analyzed in several reports. They all concluded in favour of buses; the 1974 report on *The Future of the Tramway* reached this verdict from a total assessment of the impacts on urban development, transport infrastructure, environment, and economy. When the report was discussed in the City Council in 1975, there were three main options: extension, *status quo* and close-down. This is the kind of choice set that generates single-peaked individual rankings. Neither the extreme pro-tram nor the con-tram block was in majority, and at least since Black and Newing (1951) it has been known that if we confine ourselves to a single issue dimension, the outcome will be the median on that dimension. So one would expect the compromise of *status quo* to be chosen, which was indeed the result, and it marks the transition from phase (a) to phase (b).

When the City Council confirmed in December 1979 that the trams on the two remaining lines should go on running, heavy investment appeared to be necessary. In 1982 it was decided to buy 11 new articulated tramcars. The question is how the politicians came to vote for the closure of the Elgeseter–Voldsminde line in 1982 and 1983 while at the same time confirming the purchase of the tramcars. Part of the answer is that the compromise mechanism of the last paragraph was again to work. Since the reports from 1974, 1976 and 1979 had all taken a negative stand towards extensions, expanding the system was not any longer a main option, however. This gave the con-tram block the possibility of an easy victory. All they needed to do was to keep up the pressure for total close down and wait for a suitable opportunity to vote once again over the future of the tramway. The right moment came with the important investment decision in 1982. Since the 'extreme' proposals this time were *status quo* and total close-down, the likely outcome of the vote was a partial close-down. The relative strength of the political blocks had not changed much from 1975, so the result was as expected. The Elgeseter–Voldsminde line was shut down, leaving only the Lian–Lade line in operation.

For many years the Labour Party and the Socialist Left Party made up a compact pro-tram block. The votes of the representatives of both parties were necessary to ensure a majority for continued operation of the tramlines. The need for heavy investments in 1982 appears to have sown deep doubt in the pro-tram camp, however. This paragraph outlines the dilemma most peculiar to the socialists. They realized that the purchase of 11 costly tramcars would seriously limit the investments in other parts of the public transport system. All the professional reports found the operation of the tramway system to be more expensive than an equally good bus service. Moreover, the tram is only useful to a minority, as the argument goes, while the buses can serve everybody. The conviction spread in the party that an altogether better public transport system, giving most people a more frequent service, would result from substituting trams with buses. The question bothering the Socialists was whether it is right to offer people in a few districts a high quality tram service at the expense of the majority of the inhabitants. Considerations of fair distribution of public funds came to the fore, and in this perspective advocating the tram seemed to be an inequitable and thus unsocialistic policy. One can argue from an environmental point of view that more motorists will shift to public transport the better the overall service, and that transfer from car to bus is likely to reduce pollution. This notwithstanding, the environmental argument making an impact in the debate is that the trams pollute very little. Hence, the Socialists' dilemma is how to trade off fairness and environmental improvement. The new issue dimension of just vs environmentally friendly transport policy made the Socialist Left Party rearrange their preference ordering in favour of fairness. This paved the way for tramline closure and transition from phase (b) to (c).

Attention is now directed to the transition from phase (c) to phase (d) in 1988–1989. New political actors were important to this transition. It will also be shown what new input to the decision process was crucial for inserting a pro-tram attitude in most of the City Council. Of course, the con-alternative still was that there should be no trams operating in Trondheim; that is, a confirmation of the actual situation since June 1988.

The pro-tram sentiment was still strong, however, both in the City Council and among the populace. It seemed that for many years to come one would have to calculate with the possibility of proposals for reopening the tramlines as a municipal company. The following four characteristics of the political situation made the con-tram block fear that such a proposal might pass at some unfortunate moment.

(1) The Socialist Left Party still had trouble with their fairness/environment trade-off in the debate over the tram. Nevertheless, they had returned to the environmental pro-tram position before the local election in 1987.
(2) The pro-tram City List was placed in a key position in the City Council. As monomaniacs devoted to virtually nothing but the tram, they might be able to exert unpleasant pressure.

(3) The track between Lian and the city centre was still intact. The action group Friends of Gråkallbanen keenly supported the idea of a museum tramline. Shortly after the decision on the closure of the Lade–Lian line, the City Council decided to retain the track to Lian for this purpose. The rails on parts of the other lines were rapidly broken up by the County Engineer's Department. Hence, the promotion of the museum tramline in 1988—although an intermezzo in the debate—was important for the reopening alternative to remain realistic.

(4) The £10 million investment in tramcars, depot, and runway gave no return at all at the time, and this was becoming a political problem. It was not only a question of economy, but of credibility and the politicians' image as rational actors in the eyes of the constituency. A politically rational solution was just as badly needed as an economically rational one. A reopening of tram operation at a modest scale would at least put the facilities and most of the equipment to use. As unsuccessful sales promotion measures multiplied, use vs misuse of the tramcars emerged as an important new issue dimension in the collective decision process. Non-use was associated with economic loss and political incompetence. For the first time in three decades there seemed to be an economic main argument *in favour* of the tram.

So far I have shown that at least two alternatives were available in 1989 regarding the fate of the tram. As will be seen, however, there was a third option. The Friends of Gråkallbanen declared that they were willing to operate the line between Lian and the city centre on a private basis, and in so doing they generated a new alternative. To most of the representatives in the City Council this appeared to be a feasible compromise. The pro-tram block had to be content with just one line, and this line is only about half of the former Lian–Lade line. Moreover, at least formally the tram would have to manage without municipal operating subsidies. The con-tram block had to accept that subsidies were indirectly given through Gråkallbanen's hiring of tramcars, runway, and depot at a zero price. They also risked continued pressure for extension and municipal investment grants. The great relief for the con-tram block was to avoid public responsibility for tram operation, juridically speaking.

The chance of implementing the compromise was stimulated by the existence of the City List. Their voters and their representatives in the City Council were largely recruited from the political left. However, since their overriding goal was to get the tram back in operation, they could not be trusted to vote with the Labour–Socialist block if the Conservatives had something to offer them on the tram case. Such a situation occurred in the summer of 1989. Even though it was in the middle of an election period, a new mayor was to be appointed by the City Council. The Conservatives were looking for support to their candidate, and the City List offered their votes if the Conservative Party would back the reopening of a tramline. In this piece of horse trading the City List revealed false preferences—i.e. voted strategically—in one vote in order to have their way in another one. In Arrow's terms, the social choice in the tram case was clearly dependent on irrelevant alternatives.

A lot more could be said about the manoeuvres of the City List in 1989. Just a little is told here in order to exemplify how formal procedure can determine the political outcome. The administrative Chief Officer found that the Conservatives' reopening proposal would have budgetary consequences. Since it was not backed by a majority of the City Council, a second vote on the private operation of Gråkallbanen had to be arranged. There is usually no debate before such a second vote, which takes place at the beginning or at the end of a Council meeting. On this occasion the vote was scheduled at the end of the agenda. However, the representatives of the Progress Party had the standing agreement among themselves to leave the City Council meetings at 10 p.m., and most of them left before the tramline vote. The result was that the reopening proposal of the Labour Party passed this time. It is quite likely that this change was decisive, not for the reopening but for the approval of permanent operation of Gråkallbanen in 1992, since free hire of tramcars, depot, and track was included only in the proposition of the Labour Party. It is one of the many ironies of the tram case that the City List was unable to break out of the co-operation with the Conservatives, and thus voted *against* the reopening proposal most advantageous to the tram.

With the mayor-and-tram package, the Conservative Party voted in favour of the tram for the first time since the 1950s. One reason is that reopening would have been the result anyway, since the Labour–Socialist block advanced a competing reopening proposal which would have been supported by the City List if the Conservatives let them down. Another reason is more interesting in a social choice perspective, though. After all the years of political turbulence around the tram, agreement on a compromise holding out the promise of peace, of healing social and political wounds, and making it possible to allocate time and energy for other pressing problems had a high political value of its own. This mechanism was dealt with in Section 4.2 discussing properties of disjointed incrementalism serving to relax the independence condition.

The point to be added here is that compromise may be desired to such an extent that new irrationality is brought into the process. Did the Conservatives really believe that one could avoid subsidizing the tram over public budgets just because it was to be run by a private company? The municipal administration had told them that this was utterly unlikely. Was the credulity a kind of self-deception, of closing one's eyes and pretending not to know? It was known almost for certain in 1989 that to reach break even, the Gråkallbanen had to be offered very good terms, forcing the municipality to deviate from commercial renting conditions. Even at zero-price hire, the private tram company is not able to allow any sum of money for depreciation of capital. It is hence obvious that future investments will have to be heavily subsidized if the line is to remain open.

The self-deception at the reopening point in the tramline process was serious, because it arranged for a development which may turn out to be practically irreversible. Once the longed-for compromise is established and the trams are again in traffic, it will be a heavy political burden to destroy the peace and stop operations even if agreed conditions are not met. The unsuccessful attempt at close-down in 1992 demonstrated just that. Instead of closing the tramline, the City Council released the company from debt and voted for permanent operation. Against this background, the most recent development of the case is not surprising: in December 1993, a majority of the City Council voted to grant £0.2 million to prepare an extension of the tramline ending in a downtown loop. In principle, a clearance signal was given for laying new rails to replace some of those broken up in 1988. And so the cycle continues, still contributing to 'tragedy' as understood in this section.

I have not proved that the points dealt with in this section were decisive in moving the political decision process on the tram from phase to phase. I hope to have shown, though, that strategies have been at play, which might well have such effect. The political phenomena treated can be grouped as agenda control, strategic voting and manipulation of dimensions. As Riker (1986) asserts, these strategies are central to the art of political manipulation aiming to sway decisions to suit one's own interests.

NOTES

1. Thanks are due to my colleague Ragnvald Sagen for reading and commenting on the manuscript for this chapter.
2. I draw heavily on case material presented by Aanund Hylland in a lecture on 'The will of the people' at the University of Trondheim on 2 November 1993. The case is being analyzed further by Hylland, and more detailed results will, according to plan, be published by him later on.
3. Trondheim is the third biggest city in Norway and had a population of about 140,000 in 1993. There has been a tramline system in the city since 1901.
4. The historical facts for this case study were collected and systematized by stud.-techn. Trude Tørset for a student project at the Department of Transportation Engineering, Norwegian Institute of Technology.

CHAPTER 7

Conclusions

The analysis of the possibilities of meaningful social choice is momentous to planning theory. It questions whether planning as a collective choice process can be democratic and at the same time meaningful in the sense of yielding a non-arbitrary outcome. According to the impossibility theorem of Arrow it cannot. The main thrust of Arrow's theorem and the associated literature is that there is an insolvable tension between logicality and fairness. The purpose of this essay is to demonstrate how properties of well known planning modes alleviate this tension. It is suggested here that the planners can trade off logicality and fairness—unambiguity and democracy—in different ways. And they can vary the location of their efforts along the dialogue–technique scale and the synoptic–incremental scale.

Social choice studies of decision processes have their origin in economic welfare theory, but the field has gradually developed into a social science discipline in its own right. The planning procedures can in many respects be regarded as a process leading up to public decisions. The essay explores the view of planning as a technique for amalgamating individual preference orderings so that transitive collective choice results.

It seems reasonable that the society's prevailing view on how to design collective decision-making processes influences the way public planning is conducted. Since the ideal decision process does not exist, trade-offs must be made. Sometimes the important thing will be to avoid applying properties of the preference structure about which reliable knowledge cannot be obtained. Then one should observe Arrow's claim for ordinality. Alternatively, one could try to evade Arrow's impossibility theorem by developing planning processes which give ample opportunities for relaxing his fairness condition of unrestricted scope. Disjointed incrementalism works in this direction, as it stimulates concentration on a single issue dimension and therefore the formation of single-peaked preferences. The synoptic planning mode tends to relax logicality by dispensing with the independence of irrelevant alternatives, while retaining the fairness conditions. Besides, in most practical planning the violation of ordinality is accepted, and so the perfect information approach of synoptic planning is more reasonable.

The following conclusions are derived from the text:

(i) The plan-making process concluding with a recommendation to the political decision-making body can be studied as a process for making collective choice—a device for amalgamating individual preferences.

(ii) In the planning process, the number of people affected and therefore entitled to participate in the decision process, their individual preference orderings, and the choice set, cannot be taken as given. Arrow does, however, regard these variables as exogenous. Treating them as endogenous is no way to by-pass the social choice paradox, however.

(iii) Synoptic planning often relaxes the independence condition by relying not only on individual preference orderings but also on the results of policy analysis in determining whether an action should be recommended. Moreover, maximization requires information in addition to pairwise rankings.

(iv) All the evaluation techniques capable of recommending a certain alternative violate the ordinality condition. They start by treating each consequence separately and thus need to weigh them to form a total score.

(v) Some features of disjointed incrementalism promote single-peakedness. This goes for the deliberate neglect of some values and consequences, the predilection to move away from trouble rather than systematically approaching specified goals, and the delimitation of choice to a few familiar alternatives.

(vi) Disjointed incrementalism violates the pairwise-choice format (of the independence condition) by letting the *status quo* play a special part, by interweaving what is desired and what is feasible, and by striving for agreement. In addition, *ex post* reasoning combined with the mechanism of cognitive dissonance counteracts decision cycles.

(vii) In the face of the social choice paradox, i.e. when the Arrow conditions are accepted, it seems impossible to give a generally applicable, operational and meaningful content to the concept of 'the public interest'.

(viii) To the extent that planning contributes to public deliberation, civic education and political socialization, it may further social homogeneity making it less serious to relax Arrow's condition of unlimited scope.

(ix) Assisted negotiation, like mediation and facilitation, aims at dialogue, and in so doing it increases the likelihood of single-peakedness.

Arrow claims that collective decisions based only on reliable information will be either indeterminate or undemocratic. Increased credibility would then be given to the argumentation for freedom under planning if it could be demonstrated that the impossibility theorems of social choice theory do not apply to public planning processes. Alternatively, one might argue that the Arrow conditions are often unduly strict in political systems where collective decisions are assisted by extensive public planning capable of shaping public opinion. Possibly, one might be able to specify often occurring political settings in which modifications of Arrow's conditions would not give too unappetizing consequences. A final conclusion on this matter is not reached here. I have not thoroughly analyzed what relaxations of the Arrow conditions implied by the various planning modes are (1) most likely to result in transitive collective preference orderings, and (2) least likely to have intolerable political implications. These remain future tasks for research in planning theory. Other themes following the line of investigation presented here are also suggested. Firstly, further theoretical knowledge of the functions of planning might be drawn from a similar analysis based on Amartya Sen's paradox of the Paretian liberal and other impossibility theorems. Secondly, the institutional setting and the practice of planning could be studied to sort out elements contributing to unambiguity without causing intolerable unfairness.

Bibliography

Aldrich, J. (1977) The dilemma of a Paretian liberal: some consequences of Sen's theorem. *Public Choice* **30**(Summer), 1–21.

Amy, D. J. (1987) *The Politics of Environmental Mediation.* Columbia University Press, New York.

Arendt, H. (1970) *On Violence.* Harcourt, Brace & World, New York.

Arrow, K. J. (1963) *Social Choice and Individual Values.* Wiley, New York.

Black, D. (1958) *The Theory of Committees and Elections.* Cambridge University Press, Cambridge.

Black, D. and Newing, R. A. (1951) *Committee Decisions with Complementary Valuation.* William Hodge, London.

Brams, S. J. and Fishburn, P. C. (1983) *Approval Voting.* Birkhuser, Boston.

Braybrooke, D. and Lindblom, C. E. (1963) *A Strategy of Decision.* Free Press, New York.

Brubaker, R. (1984) *The Limits of Rationality. An Essay on the Social and Moral Thought of Max Weber.* George Allen & Unwin, London.

Bruzelius, N. (1979) *The Value of Travel Time.* Croom Helm, London.

Cormick, G. W. (1982) Intervention and self-determination in environmental disputes: a mediator's perspective. *Resolve* **1**(Winter), 1–7.

Diesing, P. (1950) The nature and limitations of economic rationality. *Ethics* **61**(1), 12–26.

Diesing, P. (1962) *Reason in Society: Five Types of Decisions and Their Social Conditions.* University of Illinois Press, Urbana.

Donaldson, D. and Weymark, J. A. (1988) Social choice in economic environments. *Journal of Economic Theory* **46**(2), 291–308.

Douglas, J. (1984) How actual governments cope with the paradoxes of social choice. *Comparative Politics* **17**(Oct.), 67–84.

Etzioni, A. (1967) Mixed scanning: a 'third' approach to decision-making. *Public Administration Review* **27**(Dec.), 385–392.

Faludi, A. (1986) *Critical Rationalism and Planning Methodology.* Pion, London.

Faludi, A. (1987) *A Decision-Centred View of Environmental Planning.* Pergamon Press, Oxford.

Feld, S. L. and Grofman, B. (1988) Majority rule outcomes and the structure of debate in one-issue-at-a-time decision-making. *Public Choice* **59**(3), 239–252.

Festinger, L. (1964) *Conflict, Decision and Dissonance.* Stanford University Press, Stanford.

Forester, J. (1989) *Planning in the Face of Power.* University of California Press, Berkeley.

Forester, J. (1993) *Critical Theory, Public Policy, and Planning Practice. Toward a Critical Pragmatism.* State University of New York Press, Albany.

Friedmann, J. (1973) *Retracking America. A Theory of Transactive Planning.* Anchor Press/Doubleday, Garden City, New York.

Frohock, F. M. (1987) *Rational Association.* Syracuse University Press, Syracuse, NY.

Green, D. P. and Shapiro, I. (1994) *Pathologies of Rational Choice Theory. A Critique of Applications in Political Science.* Yale University Press, New Haven.

Grofman, B. and Uhlaner, C. (1985) Metapreferences and the reasons for stability in social choice: thoughts on broadening and clarifying the debate. *Theory and Decision* **19**(1), 31–50.

Habermas, J. (1971) *Towards a Rational Society.* Heinemann, London.

Habermas, J. (1984) *The Theory of Communicative Action. Volume 1: Reason and the Rationalization of Society.* Heinemann, London.

Habermas, J. (1987) *The Theory of Communicative Action. Volume 2: Lifeworld and System: A Critique of Functionalist Reason.* Beacon Press, Boston.

Hayek, F. (1944) *The Road to Serfdom.* Routledge & Kegan Paul, London.

Heskin, A. D. (1980) Crisis and response. A historical perspective on advocacy planning. *American Planning Association Journal* **46**(1), 50–63.

Hill, M. (1968) A goals-achievement matrix for evaluating alternative plans. *Journal of the American Institute of Planners* **34**(1), 19–29.

Hylland, A. A. (1987) Er demokrati mulig? (Is democracy possible?) *Sosialøkonomen (Oslo)* **41**(9), 15–22.

Johansen, L. (1979) The bargaining society and the inefficiency of bargaining. *Kyklos* **32**(3), 497–522.

Kadane, J. B. (1972) On division of the question. *Public Choice* **13**(Fall), 47–54.

Kelly, J. S. (1988) *Social Choice Theory. An Introduction.* Springer, Berlin.

Klemisch-Ahlert, M. (1991) Independence of the status quo? A weak and a strong impossibility result for social decisions by bargaining *Journal of Economics/Zeitschrift für Nationalökonomie* **53**(1), 83–93.
Lichfield, N. (1995) *Community Impact Evaluation*. University College Press, London.
Lindblom, C. (1959) The science of 'muddling through'. *Public Administration Review* **19**(2), 79–88.
Lindblom, C. (1979) Still muddling, not yet through. *Public Administration Review* **39**(6), 517–526.
MacKay, A. F. (1980) *Arrow's Theorem. The Paradox of Social Choice: A Case Study in the Philosophy of Economics*. Yale University Press, New Haven.
McKelvey, R. D. (1976) Intransitivities in multidimensional voting models and some implications for agenda control. *Journal of Economic Theory* **12**(3), 472–482.
March, J. G. (1978) Bounded rationality, ambiguity, and the engineering of choice. *Bell Journal of Economics* **9**(2), 587–608.
Miller, D. (1992) Deliberative democracy and social choice. *Political Studies* **40,** 54–67.
Miller, N. R. (1983) Pluralism and social choice. *American Political Science Review* **77**(3), 734–747.
Niemi, R. G. (1983) Why so much stability? Another opinion *Public Choice* **41**(2), 261–270.
Pearmain, D. and Kroes, E. (1990) *Stated Preference Techniques. A Guide to Practice*. Steer Davies & Gleave, Richmond.
Pildes, R. H. and Anderson, E. S. (1990) Slinging Arrows at democracy: social choice theory, value pluralism, and democratic politics. *Columbia Law Review* **90**(8), 2121–2214.
Piven, F. F. (1970) Whom does the advocacy planner serve? *Social Policy* **1**(May/June), 32–37.
Plott, C. R. (1976) Axiomatic social choice theory: an overview and interpretation. *American Journal of Political Science* **20**(3), 511–596.
Radcliff, B. (1992) The general will and social choice theory. *Review of Politics* **54**(1), 34–49.
Rasch, B. E. (1990) Lokket av Pandoras krukke: En voteringsteoretisk analyse av Stortingets Hurum-vedtak (The lid off Pandora's box: a voting analysis of the Hurum decision in the Norwegian parliament). In *Oppstyr og styring rundt flyplass (Commotion and Control in the Airport Case)*, ed. K. Midgaard. Dreyers Forlag, Oslo.
Rawls, J. (1972) *A Theory of Justice*. Oxford University Press, Oxford.
Redekop, J. (1995) Arrow theorems in economic environments. In *Social Choice, Welfare, and Ethics*, ed. W. A. Barnett *et al*., pp. 163–185. Cambridge University Press, Cambridge.
Riker, W. H. (1982) *Liberalism Against Populism. A Confrontation Between the Theory of Democracy and the Theory of Social Choice*. W.H. Freeman, San Francisco.
Riker, W. H. (1986) *The Art of Political Manipulation*. Yale University Press, New Haven.
Sager, T. (1981) The family of goals-achievement matrix methods: respectable enough for citizen participation in planning? *Environment and Planning A* **13**(Sept.), 1151–1161.
Sager, T. (1990) *Communicate or calculate. Planning theory and social science concepts in a contingency perspective.*. Dissertation 11, Nordic Institute for Studies in Urban and Regional Planning, Stockholm.
Sager, T. (1993) Paradigms for planning: a rationality-based classification. *Planning Theory* **9,** 79–118.
Sager, T. (1994) *Communicative Planning Theory*. Avebury, London.
Sager, T. (1995) Dialogical incrementalism: competing rationalities and the cultivation of the synoptic/incremental distinction. *Journal of Architectural and Planning Research* **12**(2), 121–140.
Samuelson, W. and Zeckhauser, R. (1988) Status quo bias in decision making. *Journal of Risk and Uncertainty* **1**(1), 7–59.
Schwartz, T. (1981) The universal-instability theorem. *Public Choice* **37**(3), 487–501.
Schwartz, T. (1986) *The Logic of Collective Choice*. Columbia University Press, New York.
Seabright, P. (1989) Social choice and social theories. *Philosophy and Public Affairs* **18**(4), 365–387.
Sen, A. (1970) *Collective Choice and Social Welfare*. Holden-Day, San Francisco.
Sen, A. (1983) Liberty and social choice. *Journal of Philosophy* **80**(1), 5–28.
Sen, A. (1985) Social choice and justice: a review article. *Journal of Economic Literature* **23**(4), 1764–1776.
Shepsle, K. A. (1979) Institutional arrangements and equilibrium in multidimensional voting models. *American Journal of Political Science* **23**(Feb.), 27–59.
Shepsle, K. A. and Weingast, B. R. (1981) Structure-induced equilibrium and legislative choice. *Public Choice* **37**(3), 503–519.
Strasnick, S. (1976) The problem of social choice: Arrow to Rawls. *Philosophy and Public Affairs* **5**(3), 241–273.
Susskind, L. and Cruikshank, J. (1987) *Breaking the Impasse. Consensual Approaches to Resolving Public Disputes*. Basic Books, New York.
Tanguiane, A. S. (1994) Arrow's paradox and mathematical theory of democracy. *Social Choice and Welfare* **11**(1), 1–82.
Uhlaner, C. J. (1989) 'Relational goods' and participation: incorporating sociability into a theory of rational action. *Public Choice* **62**(3), 253–285.
Weiss, A. and Woodhouse, E. (1992) Reframing incrementalism: a constructive response to the critics. *Policy Sciences* **25**(3), 255–273.
Young, H. P. (1974) An axiomatization of Borda's rule. *Journal of Economic Theory* **9**(1), 43–52.

FSC
www.fsc.org
MIX
Papier | Fördert
gute Waldnutzung
FSC® C083411

Kolibri 360 GmbH
Ferdinand-Jühlke-Straße 7
99095 Erfurt, Deutschland
produktsicherheit@kolibri360.de